Nanocomposites

SCIENCES

Mechanics, Field Director – Gilles Pijaudier-Cabot

Structural and Solid Mechanics, Subject Head – Julien Yvonnet

Nanocomposites

Preparation, Characterization and Modeling

Coordinated by

Jinbo Bai

iSTE

WILEY

First published 2022 in Great Britain and the United States by ISTE Ltd and John Wiley & Sons, Inc.

ISTE Ltd
27-37 St George's Road
London SW19 4EU
UK

www.iste.co.uk

John Wiley & Sons, Inc.
111 River Street
Hoboken, NJ 07030
USA

www.wiley.com

Library of Congress Control Number: 2021951379

British Library Cataloguing-in-Publication Data
A CIP record for this book is available from the British Library
ISBN 978-1-78945-031-6

ERC code:
PE3 Condensed Matter Physics
PE3_10 Nanophysics: nanoelectronics, nanophotonics, nanomagnetism, nanoelectromechanics, etc

Contents

Foreword

Polymer Nanocomposites: Are There Scientific Questions with No Answers?

Jean-François GÉRARD
INSA, Lyon, France

Nanotechnologies have now become part of our daily lives. Scientific and technological developments dedicated to nano-objects, their manipulation and their integration into nanoscale devices have been the subject of intense activity in recent years. These achievements make it possible for some of them to reach an unprecedented maturity. Considering the nanometric scale allows us to push back many limits that were previously assumed to be insurmountable, due to the fact that, at such a scale, the physical properties do not generally respond to conventional laws.

Thus, by considering new approaches, and particularly by taking nano-objects that are already widely studied into account, nanomaterials can be prepared following unconventional routes or by assembling nanometric blocks in order to provide specific functionalities. Among these nanomaterials, polymer matrix nanocomposites have been the subject of intense research and some industrial developments. Indeed, how can we not expect original synergies that lead to the design of innovative compounds by combining the immense versatility of the physical behaviors of polymers with the great diversity offered by the nano-objects that can be associated with them: 1D to 3D nano-objects, modular interfacial interactions (covalent, Van der Waals, ionic bonds), various manufacturing routes, etc.?

This diversity surely explains the wealth of scientific literature that has been dedicated to nanocomposites over the last two decades. Despite the fact that all polymers have been associated with all types of nano-objects (nanoparticles, nanotubes, nanosheets, etc.), we should acknowledge that relatively few works contribute to our understanding of the elementary phenomena that govern the behavior of polymer matrix nanocomposites. This understanding remains fundamental in developing appropriate approaches that will lead to technological breakthroughs that are deployable on a large scale.

It is, therefore, essential that we work at the crossroads of scientific disciplines, namely chemistry, physical chemistry and physics, nano-/micromechanics of interfaces, modeling the physical properties at all scales (mechanical, electrical/dielectric, transport, etc.), and so on. Of course, in order to propose a pertinent approach, a scientific approach consisting of disciplinary knowledge is required, but their combination is essential to address such a multidisciplinary subject in the most relevant way. Indeed, the field of polymer matrix nanocomposites is quite emblematic since it absolutely requires a sharing of expertise at the borders between disciplines.

This book is a perfect illustration of this cross-fertilization of knowledge and the collaboration between researchers from different backgrounds in order to question and propose new morphologies of polymer nanocomposites, design or modeling approaches, and so on. Therefore, it is with curiosity that we uncover this collection of contributions from enthusiastic researchers who aim to respond to this significant challenge of providing scientific answers that have not yet been documented on the physico-chemical/morphological/physical properties of polymer matrix nanocomposites.

1

Graphite and Graphene Nanoplatelets (GNP) Filled Polymer Matrix Nanocomposites

Marc PONÇOT, Adrien LETOFFE, Stéphane CUYNET, Sébastien FONTANA and Lucie SPEYER
Institut Jean Lamour, University of Lorraine, Nancy, France

1.1. General information on graphene

1.1.1. *Definition and structure*

Graphene is a plane of carbon atoms in sp^2 hybridization. It is the basic constituent of graphite, carbon nanotubes and fullerenes. Figure 1.1 shows a graphene plane.

Graphene can present some structural defects, although its low dimensionality reduces the number of possible imperfections. For instance, the graphene plane may not be strictly composed of hexagons of carbon atoms: these are Stone–Wales defects, where four hexagons are transformed into two pentagons and two heptagons when one of the bonds rotates. Single and double gaps can appear, as well as dislocations, atoms and impurities of substitution. These defects have a strong influence on the physical properties of graphene. In addition, some defects generate dangling bonds that increase the chemical reactivity of graphene (Banhart *et al.* 2011).

Nanocomposites,
coordinated by Jinbo BAI.

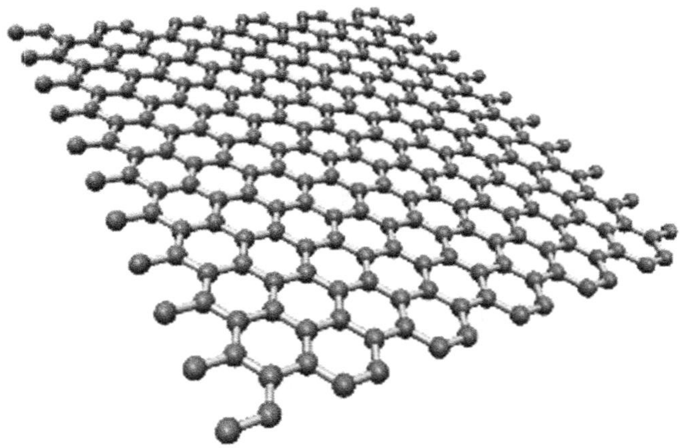

Figure 1.1. *Representation of a graphene plane*

1.1.2. *Structures associated with graphene*

As will be discussed later in this chapter, a single graphene plane is difficult to obtain and synthesis methods most often result in the stacking of multiple planes, often referred to in the literature as graphene itself. In fact, a specific terminology for carbonaceous materials with a two-dimensional character has been established (Bianco *et al.* 2013) and the following structures should be distinguished from the single graphene plane:

– *double-wall or triple-wall graphene*: structure in free suspension or resting on a substrate, composed of a stack of two or three graphene planes of extended lateral dimension;

– *few-layer graphene (FLG)*: structure in free suspension or resting on a substrate for a stack of two to five planes;

– *multi-layer graphene (MLG)*: structure in free suspension or resting on a substrate for a stack of six to 10 planes;

– *graphite nanosheets*: two-dimensional graphitic materials with one dimension not exceeding 100 nm;

– *nanoribbon graphene*: graphene plane as defined above, with lateral dimensions not exceeding 100 nm and 10 nm, respectively;

– *graphenic carbonaceous materials*: include all of the carbonaceous materials made up of sp^2-hybridized atoms, like the structures defined above, or the disordered carbons (carbon blacks, activated carbons) and graphite.

1.1.3. *Graphene properties*

The calculated and measured properties of graphene are unique and remarkable, which explains the interest of the scientific community. Wallace (1947) showed the particular electronic structure of graphene. Thus, graphene is considered as a zero-gap semiconductor. The mobility of charge carriers is high and has been measured up to 230,000 $m^2 \cdot V^{-1} \cdot s^{-1}$ (Bolotin *et al.* 2008). This value corresponds to the highest value for a semiconductor. Its thermal conductivity is also remarkable. It reaches 5,300 $W \cdot m^{-1} \cdot K^{-1}$ (Rutter *et al.* 2007), which corresponds to the highest value for a material. In comparison, diamond, which is well known for its ability to evacuate heat, has a thermal conductivity of only 2,000 $W \cdot m^{-1} \cdot K^{-1}$. Its network of sp^2-hybridization atoms gives graphene a unique set of mechanical properties. The combination of a hardness equivalent to that of diamond, close to 1 TPa, a very high flexural strength, a high specific stress at break of 48,000 $kN \cdot m \cdot kg^{-1}$ and its elongation at break of 25% makes it unique (Lee *et al.* 2008; Zhu *et al.* 2010). In comparison, steel has a specific stress at break of 154 $kN \cdot m \cdot kg^{-1}$ (Zhu *et al.* 2010). Graphene has a transparency of 97.7% in the visible spectrum (Rutter *et al.* 2007; Zhu *et al.* 2010). Graphene is considered to be the most impermeable material in the world. This is due to the combination of its sp^2-hybridization carbon atom lattice with high electron density in its aromatic rings and its structure (covalent bonds and rupture stress) (Berry 2013). Another important property of graphene is that its theoretical specific surface area is 2,630 $m^2 \cdot g^{-1}$ (Ferrari *et al.* 2015), which is very high for a carbonaceous material.

1.2. Graphene preparation methods

Graphene is an amazing material because of the combination of exceptional properties conferred by its two-dimensional character and sp^2-hybridization lattice of carbon atoms. The properties are particularly noticeable in the case of a single plane of free single-crystal graphene. Many works have been dedicated to graphene synthesis since the research by Novoselov and Geim (2004). Several so-called major synthesis methods have been developed and allow very different samples to be obtained according to several criteria:

– the structural quality and chemical purity of the sample;

– the presentation of the sample: graphene planes on metallic substrate, SiC or other, or dispersion;

– the cost price and the industrializable nature of the method.

Figure 1.2 shows the materials obtained by the synthesis methods described in this chapter, according to their structural quality and their estimated cost for industrial production.

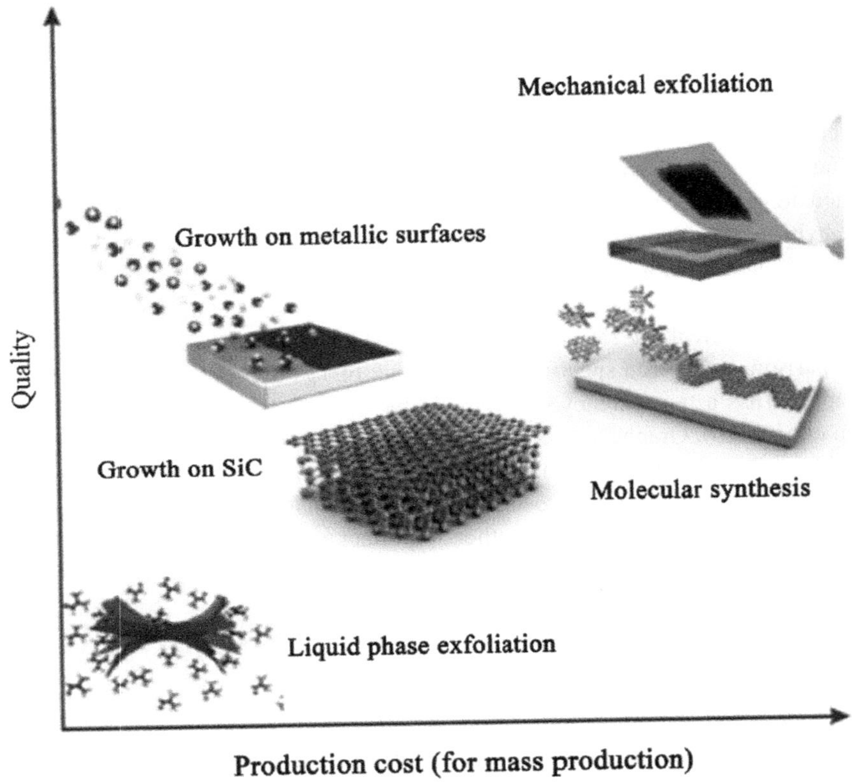

Figure 1.2. *Quality and industrial cost of graphene samples obtained by different synthesis techniques (Novoselov* et al. *2012). For a color version of this figure, see www.iste.co.uk/bai/nanocomposites.zip*

Mechanical exfoliation produces an excellent structural quality but, although very simple in principle, would generate too high a manufacturing cost. Liquid-phase exfoliation, although based on the same principle, provides graphene dispersions of lower quality, but it is much easier to obtain large quantities of material. The supported growth techniques, which will not be detailed in this chapter, are expensive techniques (energy cost, price of substrates and silicon carbide wafers, reaction chambers), but the material is of high quality, and some techniques make it possible to obtain very large surface areas (Bae *et al.* 2010). In graphite, the cohesion of the crystal along the axis $\vec{c}$ is ensured by Van der Waals bonds. The basic principle of exfoliation techniques is to break these bonds, by the so-called dry method or in liquid phase, in order to obtain graphene. It is also possible to exfoliate other graphite-based materials, such as intercalation compounds and graphite oxide (GO).

1.2.1. *Graphite exfoliation*

1.2.1.1. *Dry exfoliation*

Mechanical exfoliation is the synthesis technique made famous by Novoselov and Geim (2004). The principle is very simple and consists of applying adhesive tape on graphite (typically pyrolytic graphite because of its superior crystalline quality), peeling off the tape by isolating a certain thickness of graphene planes, and repeating the operation on the film thus obtained until the thinnest possible thickness is reached (Novoselov *et al.* 2004). The material is then usually transferred to a SiO_2/Si substrate to allow for the selection of the single walls by way of characterization techniques, such as optical microscopy. To date, mechanical exfoliation remains the technique that enables the best electronic and structural quality to be obtained. However, it is not suitable for industrial production: its main application remains basic research and the development of prototypes for various applications (Novoselov *et al.* 2012). Other dry exfoliation techniques have been developed, motivated by the crystalline quality of graphene obtained by mechanical exfoliation and in attempts to solve the problem of large-scale production. Laser ablation of graphite (Dhar *et al.* 2011) or anodic exfoliation (Shukla *et al.* 2009) have been performed and also yield high-quality material.

1.2.1.2. *Liquid-phase exfoliation by ultrasound*

Liquid-phase exfoliation of graphite is based on graphite dispersion in solvent and on the transition from dispersion to ultrasounds. Due to the hydrodynamic forces, the ultrasounds will generate a phenomenon of cavitation, namely the formation of bubbles. These bubbles will be able to break the Van der Waals bonds between the graphene planes, and the solvent must be able to stabilize the planes in suspension to prevent their reaggregation. The fundamental characteristic of the solvent to be considered in this perspective is the surface tension. Indeed, a solvent with adapted surface tension will be able to minimize the interfacial tension between the graphene sheets and the solvent molecules, and thus prevent their reaggregation. The ideal surface tension is 40–50 mJ·m^{-2}. Graphite exfoliation is thus carried out in solvents such as N-methylpyrrolidone (NMP) or dimethylformamide (DMF). A centrifugation step needs to be added to remove the thickest particles (Hernandez *et al.* 2008). This process allows dispersions of micrometer-sized particles to be obtained, generally less than 5 planes thick, with a final content of 1 wt% of single-walled graphene. A very long sonication time (more than 300 h) makes it possible to obtain concentrations of the order of 1 mg·mL^{-1} with a single-walled yield of 4 wt%, but ultrasound tends to fracture the sheets and reduce their size (Khan *et al.* 2008). Solvents such as NMP or DMF are toxic and their high boiling point (202°C for NMP and 153°C for DMF) is problematic as they evaporate slowly when the dispersion is deposited on a substrate (e.g. for the purpose of producing graphene film), which favors particle reaggregation. In contrast, isopropanol and chloroform, which have low boiling points (respectively, 82°C and 62°C), provide less concentrated graphene dispersions than NMP and DMF, possessing about 1 wt% single-layered graphene, but react less easily during a deposition process (O'Neill *et al.* 2011). Water does not present the surface tension required to stabilize graphene planes. Still, the addition of surfactants can enable the exfoliation of graphite in aqueous media, such as sodium dodecylbenzene sulfonate (SDBS) (Lotyal *et al.* 2009) and sodium cholate (Green and Hensen 2009). When graphene planes are dispersed in an aqueous medium, the hydrophobic part of the surfactant associates itself with the graphene planes and the hydrophilic part points toward the water. The whole forms a structure similar to a micelle that thus stabilizes the graphene plane. Similar to organic solvent exfoliation, dispersion by means of surfactants provides micrometer-sized particles consisting of a few graphene planes (Lotyal *et al.* 2009; Notley 2012). The continuous addition of surfactant during exfoliation makes it possible to achieve concentrations of 15 mg·mL^{-1}, and controlled separation of particles according to their thickness is possible using the density gradient centrifugation technique (Green and Hensen 2009; Notley 2012). The graphite exfoliation process, in either organic or aqueous media, does not introduce structural defects into the graphene planes. Dispersions are stable for several weeks to months.

1.2.1.3. *Graphite exfoliation and functionalization by plasma process in one step*

1.2.1.3.1. Plasma by dry-process

Environmentally friendly and economically viable synthesis processes for obtaining pure single-walled graphene with excellent physicochemical properties remain to be discovered. Nevertheless, recent research in the field of plasma physics seems promising (Dey *et al.* 2016; Levchenko *et al.* 2016; Giorcelli *et al.* 2018).

The dry plasma processes used for the preparation or the functionalization of graphene are varied. First, they are differentiated according to the gas pressure, ranging from low pressure processes (a few Pa to 100 Pa) widely known in the literature (Malesevic *et al.* 2008; Brault *et al.* 2013; Tsai *et al.* 2014; Yue *et al.* 2014) to atmospheric pressure processes in strong development (Subrahmanyam *et al.* 2009). Low-pressure plasma processes are strongly local thermodynamic non-equilibrium processes with high electron temperatures of up to a few tens of electron volts and a low ionization degree (Samukawa *et al.* 2012). Plasmas at atmospheric pressure, on the other hand, can exhibit a high degree of ionization, but low electron temperatures (Anders *et al.* 2009). The pressure difference between the two processes plays a major role in the physical and chemical action that the plasma will have on the etching material. Second, dry processes are differentiated according to the generated plasma regime, which can range from continuous discharges to pulsed discharges and even very high frequency discharges at the electron cyclotron resonance. Plasma arc and jet processes have been mentioned in the literature from the last few years (Wu *et al.* 2010a; Kim and Suh 2014), mainly because of their ease of implementation and the large quantities of material that can be processed.

1.2.1.3.2. Liquid-phase plasma

Depending on the process, plasma physics is of real interest to prepare or functionalize graphene in the dry process. It is thus possible, for example, to produce liquid-phase graphene nanosheets from asphalt by plasma (Li *et al.* 2013). On the other hand, the possibility of carrying out both production and functionalization in a single step by means of plasma process requires the removal of scientific and technical barriers, in particular by dissociating purely physical processes from reaction processes. Within this perspective, a first milestone is being achieved at the Institut Jean Lamour in Nancy, by proposing an alternative process based on plasma exfoliation and functionalization of carbon charges in aqueous phase and at atmospheric pressure. In addition to the objective of combining the production of graphene and its functionalization, this new aqueous phase process should also make it possible to avoid the effects of agglomeration and curling of graphene sheets, which are found experimentally and systematically after a synthesis process, before the one enabling the functionalization of graphene. A diagram of the experimental device is proposed in Figure 1.3.

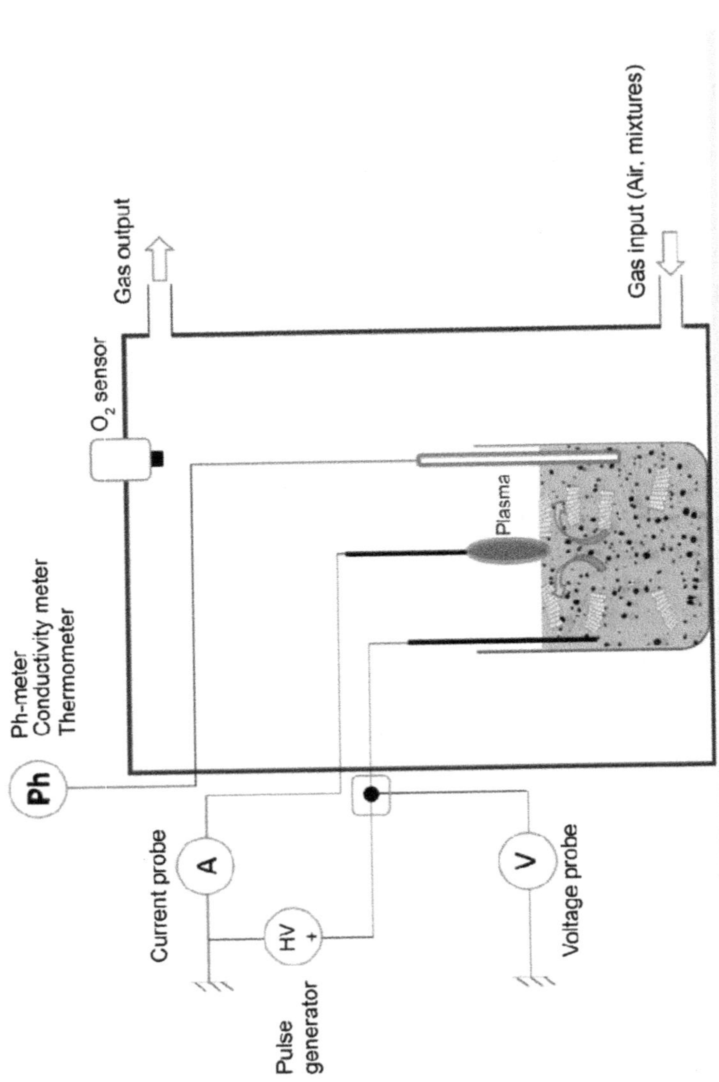

Figure 1.3. *Schematic of the plasma device allowing both the exfoliation and the functionalization of graphitic charges in solution. For a color version of this figure, see www.iste.co.uk/bai/nanocomposites.zip*

This process operates in the nano- and microsecond-pulsed arc plasma regime on the surface of an aqueous solution loaded with pure graphite particles. A voltage generator coupled to a nanosecond voltage chopper provides a square-wave electrical power signal over the range 0–30 kV and 0–100 kHz, with a pulse duration ranging from 10 to 2000 ns. An electrode immersed in the liquid is connected to the high voltage, so that potential changes are transferred to the surface of the aqueous phase. The upper electrode of the same nature is connected to the ground of the system. The plasma arc discharge is then located between the surface of the solution and the upper electrode. The initiation of the plasma, which is called breakdown, occurs under experimental conditions verifying Paschen's laws, that is to say when the product between the gas pressure and the distance between the electrodes minimizes the breakdown voltage of the gas. The interest of the high-voltage pulse regime is to promote the exfoliation process of graphite by the ignition of shock waves during plasma breakdown in the air at atmospheric pressure: the carbon planes of the carbon charges located on the surface of the liquid can then undergo a mechanical release. Studies of propagation waves originating from gas bubble formation in a similar case of discharge in liquids indicate that pressures above 10^7 Pa can be reached locally (Hamdan *et al.* 2013a). The other advantage of this discharge regime is that it allows for high energy plasma functionalization while having a low energy process, with an average power between 10 and 120 W, depending on the discharge conditions. By imposing only the gas breakdown voltage, it is possible to evolve the process to reach currents of a few amperes. To this end, the aqueous solution will undergo radical changes at the level of its initial intrinsic properties, particularly at its composition level, which will modify its pH and conductivity. This change in liquid composition is due to plasma–liquid interactions, and more specifically to the progressive solvation of plasma-derived species at the liquid surface, such as electrons, ions, and other radicals or energetic species (Lukes *et al.* 2014; Bruggeman *et al.* 2016). The evolution in the composition of the medium, its acidification and its progressive rise in temperature are all variable parameters that also promote graphite exfoliation processes on the surface and in the volume of the liquid, as well as reactional mechanisms to functionalize the graphene planes.

1.2.2. *Graphite-derived compounds exfoliation*

1.2.2.1. *Graphite intercalation compounds exfoliation*

1.2.2.1.1. Graphite intercalation compounds

Due to its lamellar structure, graphite can accommodate chemical species, called inserts, in its interplanar spaces to give graphite intercalation compounds. Intercalation is a redox reaction that is accompanied by a charge transfer between the graphene planes and the insert; graphite, being amphoteric, can react with electron acceptors and donors. In the well-known example of the KC_8 compound and of alkali metal intercalation in general, the metal acts as an electron donor. It acquires a positive partial charge, while the graphene planes have a negative partial charge.

1.2.2.1.2. Graphite intercalation compounds thermal exfoliation

Thermal exfoliation is most often carried out using intercalation compounds obtained with nitric acid and sulfuric acid. When a violent thermal shock is applied to the intercalation compound, the insert does not have time to diffuse out of the structure and vaporizes directly in the core of the material, causing the graphene planes to spread apart along the axis $\vec{c}$, due to strong pressure, and a strong increase in volume. The material thus obtained is called exfoliated graphite. It is not very dense and presents an accordion-like structure (Figure 1.4) (Furdin 1998; Inagaki *et al.* 2004; Afanasov *et al.* 2010).

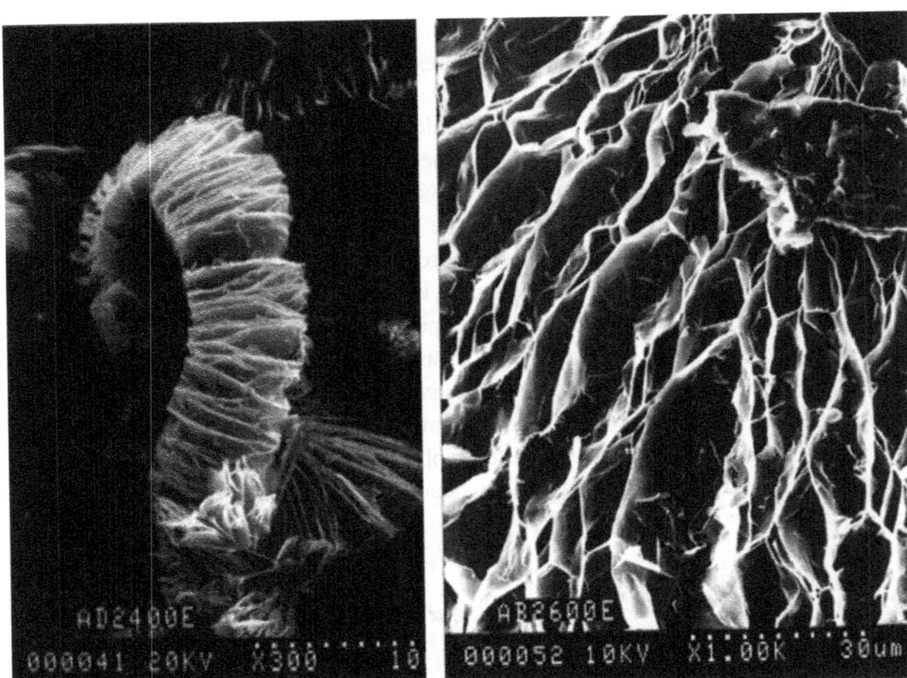

Figure 1.4. *SEM micrographs at different magnifications of exfoliated graphite developed at the Jean Lamour Institute*

Exfoliated graphite can also undergo ultrasonication and grinding steps to yield a material called flat micron graphite (FGM), consisting of single-crystal particles tens of micrometers wide and a few dozen graphene planes thick (Furdin *et al.* 1991).

1.2.2.1.3. Liquid phase exfoliation of graphite intercalation compounds

The dispersion of a graphite intercalation compound, usually based on alkali metals, in a polar solvent leads to its exfoliation due to charge separation. Electrostatic interactions stabilize the negatively charged graphene planes in suspension. The most widely used compound is KC8 (Catheline *et al.* 2011, 2012). $K(THF)C_{24}$ exfoliation has also been reported (Vallés *et al.* 2008). The exfoliations were initially performed in high-boiling solvents such as NMP but, as with graphite exfoliation, these solvents can be replaced by more easily removed solvents such as tetrahydrofuran (THF). The dispersions are stirred and centrifuged and the resulting particles, of micro-metric size, consist of a stack of one to a few graphene planes (Vallés *et al.* 2008; Catheline *et al.* 2011, 2012). The concentrations obtained are of the same order as those of dispersions obtained directly from graphite. Upon the deposition of the dispersions on a substrate, the exposure of the graphene sheets to air returns them to their 0-oxidation state and leads the $K^{\delta+}$ counterions to oxidize. The oxides are easily removed through washing. The main drawback of this method for obtaining graphene is the extreme sensitivity of the alkali metals and graphite intercalation compounds to air, which must be handled under an inert atmosphere.

1.2.2.2. *GO exfoliation*

1.2.2.2.1. Graphite oxide

GO is a compound that has been known for many years. This is graphite that has been treated in an oxidizing acid medium. Several historical methods of GO synthesis exist, such as the Hummers method or the Staudenmaier method. The Hummers method consists of treating graphite with sulfuric acid, potassium permanganate and sodium nitrate (Hummers and Offeman 1958). GO, a light-yellow solid, is thus made up of a stack of "graphene" planes carrying various types of oxygenated functions. The oxygenated functions that a carbonaceous material can exhibit are listed in Figure 1.5.

The structure of GO depends on the oxidative treatment (Botas *et al.* 2013) and the crystalline quality of the starting graphite (Botas *et al.* 2012). Therefore, it is difficult to establish a typical structure for this type of material and several models have been established (Dreyer *et al.* 2010). GO mostly contains phenol, carbonyl, carboxyl, ether and lactone functions (Figure 1.5) (Gao *et al.* 2009). These functions have the effect of doubling the interplanar distance of the compound compared to graphite (Schniepp *et al.* 2006), and oxidative treatment also leads to the degradation of the sp^2 lattice.

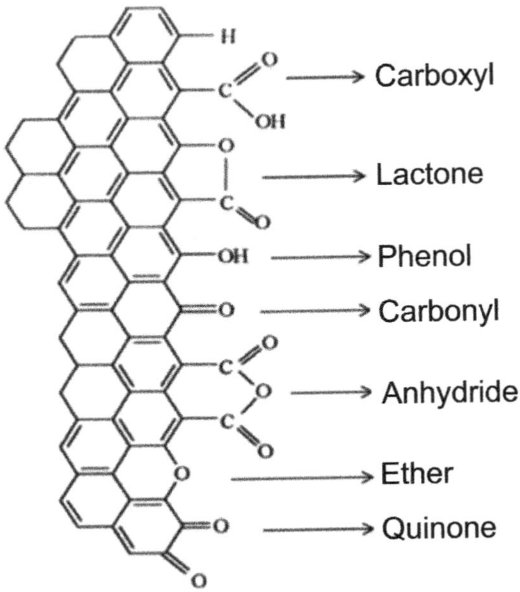

Figure 1.5. *Oxygenated functions carried by carbonaceous materials (Figueiredo* et al. *1999)*

1.2.2.2.2. GO thermal exfoliation

This technique experimented by Boehm *et al.* (1962) lead to the first probable observation of graphene and has been recently reutilized with the works on graphene synthesis. When GO is heated rapidly to a temperature of about 1,000°C, the organic functions decompose into CO and CO_2. The pressure thus generated is enough to exfoliate the graphene planes. A black solid composed of crumpled graphene sheets is obtained and after dispersion in NMP or DMF, single-walled graphene with micrometric lateral dimensions can be isolated (Schniepp *et al.* 2006; McAllister *et al.* 2007). Oxidative treatment disrupts the sp^2 lattice of the sheets and thus compromises their properties. Although thermal decomposition of the functions removes much of the oxygen, an ideal graphene structure cannot be restored. Violent heating leads to the additional formation of carbonyl and epoxy groupings, which are thermodynamically very stable and therefore difficult to remove. Furthermore, the removal of the functions leaves sp^3-hybridized atoms and atomic gaps within the planes (see Figure 1.6).

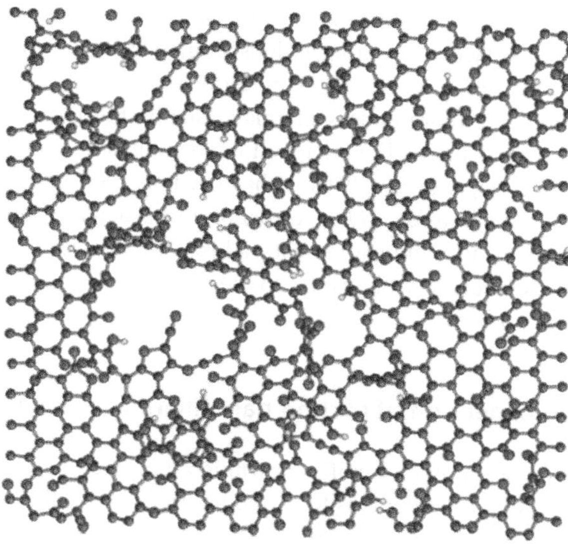

Figure 1.6. *Structure of graphene oxide after thermal exfoliation (carbon atoms in gray, oxygen in red and hydrogen in white) (Bagri* et al. *2010). For a color version of this figure, see www.iste.co.uk/bai/nanocomposites.zip*

1.2.2.2.3. Liquid-phase exfoliation of GO

The oxygenated functions grafted onto the GO planes make them hydrophilic, unlike a purely carbonaceous surface, which is highly hydrophobic. Thus, a simple dispersion in water is sufficient to exfoliate the GO and stabilize particles mostly composed of a single plane (Gomez-Navarro *et al.* 2007; Stankovich *et al.* 2007). GO also exfoliates in organic media (Traina and Pegoretti 2012). The material thus obtained is called graphene oxide. It carries the same functions as the parent GO and the carbon/oxygen atomic ratio is between 2 and 3 (Stankovich *et al.* 2007; Paredes *et al.* 2008). The characterization of the transport properties indicates a strong degradation of the sp^2 lattice: graphene oxide is a quasi-insulating material (Gomez-Navarro *et al.* 2007). In order to restore the properties of graphene oxide, a reduction step is implemented. The most commonly used reducer is hydrazine N2H4. This treatment allows a good part of the oxygen to be eliminated (the carbon/oxygen atomic ratio can rise up to 10) and significantly improves the conduction properties (Gomez-Navarro *et al.* 2007; Stankovich *et al.* 2007). However, the charge carrier mobility is 30 $cm^2 \cdot V^{-1} \cdot s^{-1}$, which still remains much lower than the results obtained with "pure" graphene (Gomez-Navarro *et al.* 2007). Moreover, as the functionalization is weaker after reduction, the particles become less hydrophilic, tend to reaggregate in aqueous medium and are very difficult to redisperse (Stankovich *et al.* 2007). Since hydrazine is toxic, several studies have been conducted to replace it. Non-hazardous compounds such as ascorbic acid (vitamin C) are equally effective (Fernandez-Merino *et al.* 2010). Furthermore, since oxygenated functions are more or less thermally stable, the reduction of graphene oxide can be achieved by thermal reduction at temperatures on the order of 1,000°C (Becerril *et al.* 2008; Bagri *et al.* 2010). It has been shown that thermal reduction could lead, at high temperatures, to the formation of particularly stable oxygenated groups that are difficult to eliminate, but it yields good or even better results than chemical reduction, especially with respect to the restoration of the sp^2-hybridization lattice and transport properties.

1.3. Methods of dispersion of carbon nanofillers in a polymer matrix

The method of dispersion of nanofillers within a polymer matrix is as impactful on the final properties of the composite as the choice of the filler and its potential chemical treatment: it can strongly modify the different percolation thresholds. Numerous methods have been developed according to the polymer matrix used and the targeted properties. These methods, for the most part, are divided into three categories: preparations by *in situ* polymerization, solution mixing and melt mixing. Moreover, each of these methods is divided into those known as non-covalent, with a nanofiller simply dispersed in the matrix, and those known as covalent, allowing the creation of chemical bonds between the filler and the matrix.

1.3.1. In situ *polymerization*

In situ polymerization dispersion is a method of nano-composite preparation that can be adapted to a wide range of thermoplastic matrices, such as polyolefins (Fim *et al.* 2010), polyamides (PAs) (Pan *et al.* 2000), polystyrenes (PSs) (Chen *et al.* 2003b), and polymethyl methacrylate (PMMA) (Chen *et al.* 2003a), but also within thermosetting matrices such as epoxy resins (Miller *et al.* 2010; Kernina *et al.* 2019). This method allows for the production of composites from matrices that cannot be used in melt mixing (thermal instability) or dispersed in solution (insoluble polymer). The method is classically carried out in two stages: the first being the mixing of one or more monomers, in powder form or dispersed in a solvent, with the nanofiller, to obtain a homogeneous mixture, then the polymerization of the monomer(s) is carried out to obtain the final nanocomposite. For example, Cromer *et al.* (2015) performed *in situ* polymerization of a polypropylene (PP) in the presence of exfoliated graphene fillers. The materials produced by *in situ* polymerization show better mechanical properties than an equivalent nanocomposite produced by melt mixing, with notably higher elongation at break. Their work highlights the impact that nanofillers can have on polymerization, and the need to modify the conditions in order to maintain the efficient polymerization reaction.

Pol'shchikov *et al.* (2013) compared two nanocomposites obtained by *in situ* polymerization via two methods. The nanofillers employed were GNPs produced from oxidized graphite according to the previously mentioned Hummers method. With their first method, the nanofiller, after being heated to 200°C, is placed inside the reactor with dissolved propylene and the metallocene catalyst. With the second method, the nanofiller is first dispersed in toluene and ultrasonicated for 10 or 30 minutes. It is then added to the reactor with propylene and metallocene catalyst. For both methods, the resulting materials were rinsed with a mixture of ethanol and HCl, then washed with alcohol and dried. In both cases, infrared spectroscopy analysis showed that the stereoregularity of the obtained iPP was not influenced by the addition of the nanofiller, but that the use of ultrasound has greatly influenced the dispersion of carbon within the matrix. The two ranges of materials show a close percolation threshold (between 2 and 3 vol%), but the ultrasonically treated materials show higher electrical conductivity and a greater impact of the filler on the mechanical properties with better elongation at break and on the crystallization of the material, with a higher crystallization temperature (Pol'shchikov *et al.* 2013).

As introduced earlier, it is possible to produce covalent bonds between the nanofiller and the matrix during the polymerization. For example, Huang *et al.* (2010) performed *in situ* polymerization of an oxidized PP/graphene nanocomposite with the creation of covalent bonds. The elaboration starts with the preparation of nanofillers via the use of Grignard reagents, in order to obtain a GO nanofiller grafted with Ziegler–Natta catalysts, called TiCl4/(BuMgCl/GO). As presented in Figure 1.7, the obtained filler is then employed for the production of a PP/GO nanocomposite by *in situ* polymerization. The final nanocomposite shows covalent bonds between the PP polymer matrix and the GO filler. These covalent bonds allow the possible incompatibility between the nanofiller to be bypassed, here polar (GO), and the non-polar polymer matrix, PP, resulting in a better exfoliationand dispersion of the nanofiller. In the work by Huang *et al.* (2010), an excellent exfoliation of GO fillers is achieved.

It should be noted that the *in situ* polymerization method can produce a very high level of exfoliation of the carbon planes, especially for intercalated graphite (ICG) or expanded graphite (EG) fillers, by the intercalation of monomer molecules between the carbon planes, and their separation during the polymerization step. This method, which can be called intercalation polymerization, is very similar to the methods used for polymer/clay composites (Ray *et al.* 2003). The graphene planes are intercalated with monomer molecules in addition to alkali metals in the case of ICGs, or between the accordion planes of EGs (Shioyama *et al.* 2000). In the case of

GO fillers, the distance between oxidized graphene planes, between 0.6 and 0.8 nm, compared to 0.34 nm for a graphite filler, also allows the intercalation of monomer molecules or polymer chains (Jang *et al.* 2008), particularly in the case of polar polymers such as polyvynilalcohol (PVA) (Ma *et al.* 2016). The major problem with *in situ* polymerization development is the complexity of its implementation, but also its high energy consumption, which makes it difficult to apply to high-volume production (Potts *et al.* 2011; Gulrez *et al.* 2013).

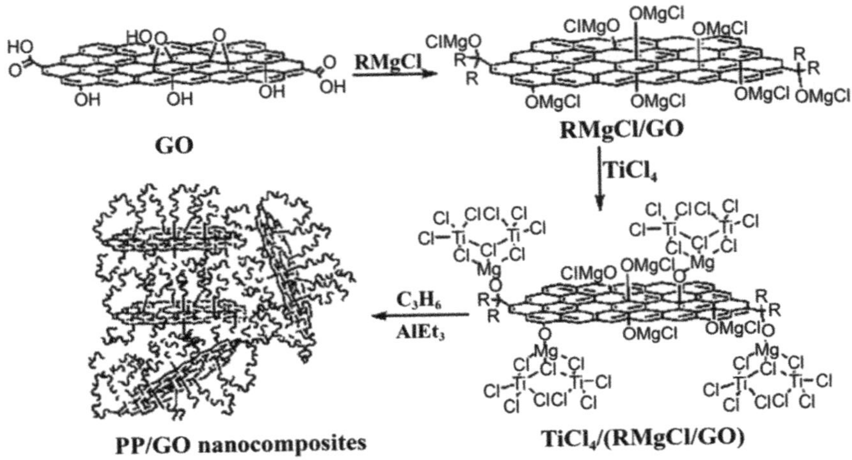

Figure 1.7. *Production of PP/GO nanocomposite by* in situ *Ziegler–Natta polymerization (Huang* et al. *2010)*

1.3.2. *Intercalation in solution*

Solution intercalation processing, like *in situ* polymerization, provides excellent results for dispersion and exfoliation of a carbon filler within a polymer matrix. The method is applicable to any matrix that can be put into solution in a solvent, such as polyethylenes (PE) (Pang *et al.* 2010) or PS (Fang *et al.* 2009). The method simply consists of dispersing the polymer and the nanofiller within the same solvent, and taking advantage of the low viscosity of the mixture to ensure its homogeneity before extracting the solvent. The method is particularly suitable for developing nanocomposites employing an oxidized graphite nanofiller and a water-soluble polymer matrix. Indeed, due to their functionalizations, GO fillers are easily dispersible in water, often by a sonication operation enabling their exfoliation. For instance, Salavagione *et al.* (2009) produced a PVA/reduced oxidized graphite nanocomposite (RGO often obtained by adding hydrazine) by solution intercalation. The carbon filler employed is an oxidized graphite (GO) developed by the Hummers method. The same method without the reduction step is also performed for the production of PVA/graphite and reduced PVA/oxidized graphite, in order to compare their exfoliation state. Thus, the graphite-filled composites show significant agglomerations in the matrix, while GO and RGO composites show better exfoliation of their fillers, especially for PVA/RGO, with an electrical percolation threshold at 1wt%. The choice of the filler influences the quality of the exfoliation and the dispersion of the final composite, and shows the necessity to adapt the solvent. The importance of the initial dispersion of the filler by ultrasound can also be seen in this case. In an optimal filler/matrix/solvent configuration, it is thus possible to obtain excellent dispersion results, allowing the percolation threshold of the produced nanocomposite to decrease. In some cases it is possible to obtain a percolation threshold lower than 0.5 wt%, as shown by Pang *et al.* (2010). In their work, a nanocomposite is developed from a high-density polyethylene (HDPE) and an EG nanofiller produced by the Hummers method. Carbon-coated HDPE particles are obtained and shaped by hot pressing. The method of development allows the creation of a percolated lattice, even at low carbon levels during forming. This result is also visible with the evolution of the conductivity represented in Figure 1.8, with an important increase between the materials at 0.005 vol% and 0.076 vol%, and a percolation threshold estimated at 0.070 vol%.

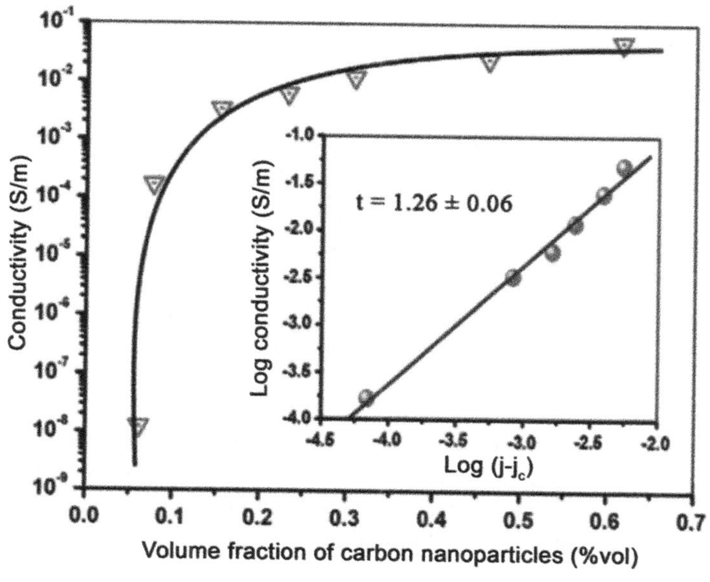

Figure 1.8. *Evolution of electrical conductivity as a function of GNP volume fraction of an HDPE matrix nanocomposite (Pang* et al. *2010). For a color version of this figure, see www.iste.co.uk/bai/nanocomposites.zip*

The main limitation of the solution intercalation method comes from the limited number of solvents that can be used, but also from the polymer matrix, which must have a high solubility in order to enable a homogeneous mixing. Moreover, in the case of certain polymers which are difficult to solubilize, such as PP, the use of polar solvents such as xylene is necessary. The intercalation method in solution then requires the evaporation of large quantities of organic solvent, which makes the method not environmentally friendly and hardly adaptable to the industrial production of nanocomposites (Bafana *et al.* 2017).

1.3.3. *Melt mixing*

Melt mixing is the most widespread method for the preparation of nanocomposites because of its ease of implementation and the large quantity of materials that can be produced. Moreover, it is the most interesting method from an ecological point of view because it does not involve solvents. Therefore, melt mixing is mainly employed by the industrial sector. There are two main technological routes for this production method: by internal mixing, where the matrix is melted before the addition of the nanofiller, and by extrusion, where the filler can be introduced either at the extrusion inlet with the polymer pellets or during the extrusion process. Because of the great adaptability of an extruder by changing the profile of the conveying screws and the possibility of varying the temperature during mixing, extrusion is the most often favored technology.

Similar to the solution intercalation method which can only be applied to matrices that are soluble in a solvent, the melt mixing method can only be used for matrices that are not too sensitive to thermal degradation. Despite this limitation, it is applicable to many matrices commonly used in industry, such as PP (Kalaitzidou *et al.* 2007a), PET (Zhang *et al.* 2010), PA (Canales *et al.* 2015) and PS (Oxfall *et al.* 2013). A major shortcoming of this method is a generally higher percolation threshold than with previous methods, mainly due to the fact that the mixture viscosity remains high.

Many authors have proposed methods to improve the dispersion of the filler within the matrix in order to improve its electrical and mechanical properties. The first development parameter that can be easily modified is the mixing speed by screw rotation. Khanam *et al.* (2016) highlight its importance in the development of a low-density polyethylene/GNP composite. Their results show an improvement in the mechanical properties and electrical conductivity of the nanocomposites, with increasing screw rotation speed due to better dispersion of the nanofiller. However, the authors note that the improvement of the properties is limited when high carbon concentrations are used. Another interesting approach is, as in the case of *in situ* polymerization or solution intercalation, the use of ultrasound during the production to allow for a better exfoliation and dispersion of the nanofillers. These can either be used during the preparation of the production to ensure the homogeneous mixing of the polymer in powder form with the filler, or directly during the extrusion mixing. He *et al.* (2018) produced nanocomposites with a GNP mass fraction between 9 and 15 wt% by a single screw extruder coupled with an ultrasonic system, as shown in Figure 1.9. Different configurations were investigated by varying the rotation speed of the extruder and the applied power of the ultrasonic treatment.

These same authors first note an increase in the number and size of the aggregates within the matrix with the increase in carbon concentration. Following the use of ultrasound, an improvement of the dispersion is observed for each concentration. Another important observation is that increasing the power of the ultrasonic treatment makes it possible to reach a better dispersion and even enables the exfoliation of the filler from a certain power, 300 W in the case of their work. The electrical conductivity of the different materials is improved with the increase in the carbon concentration, but also in the case of materials with and without ultrasound treatment. A last interesting result observed by the authors is that an increase in the extruder rotation speed during the ultrasonic treatment decreases the filler dispersion.

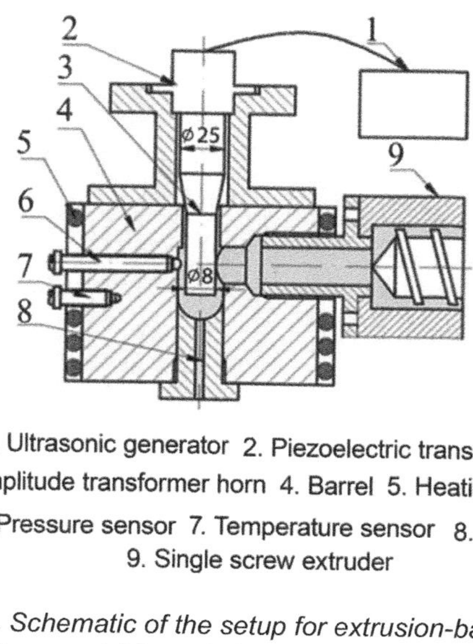

Figure 1.9. *Schematic of the setup for extrusion-based production using ultrasound (He* et al. *2018). For a color version of this figure, see www.iste.co.uk/bai/nanocomposites.zip*

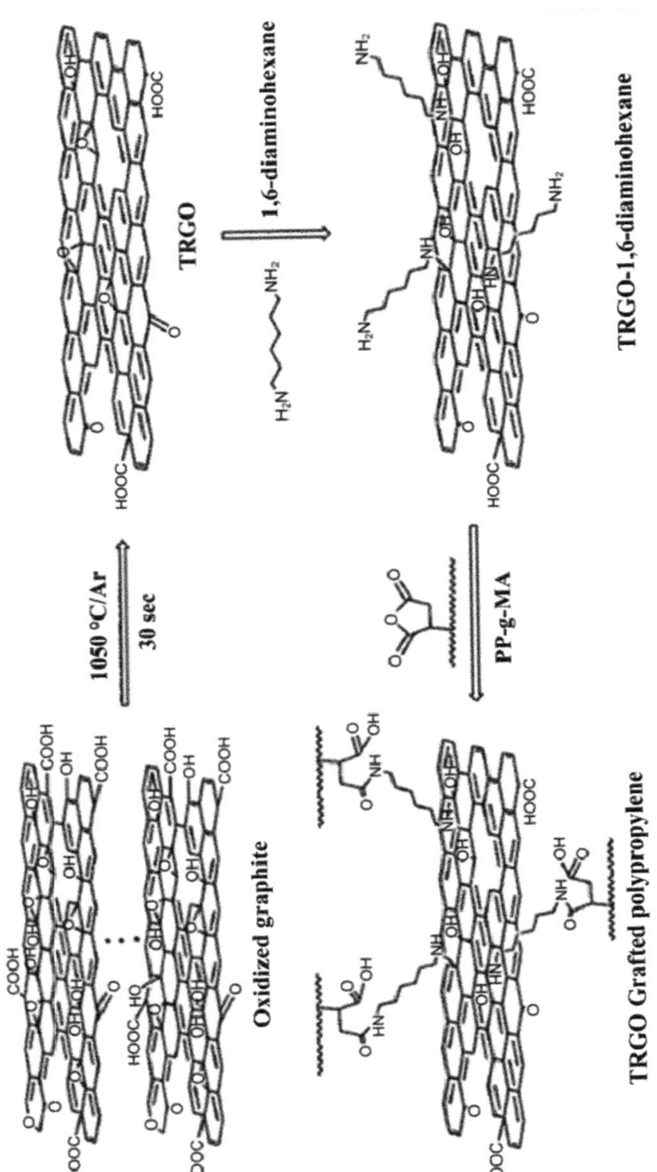

Figure 1.10. *Preparation of TRGO-grafted polypropylene (Hsiao et al. 2010)*

A classical variation of the melt mixing preparation makes use of an intermediate step called *masterbatch*. The method consists of achieving a first polymer/filler mixture with high concentrations (which can exceed 75wt% in nanofiller), then diluting this masterbatch during extrusion to reach the targeted filler concentration. The masterbatch method is indicated for the realization of nanocomposites based on a matrix with a low compatibility with the nanofiller, such as PP, because it is possible to use a derivative close to the matrix during the production of the mixture with high carbon content. For example, Rosehr *et al.* (2018) employ an atactic PP (aPP) in their works to facilitate the dispersion of their graphite nanofillers within an iPP matrix. Two mixtures were produced by solution mixing in 50 and 80 wt% carbon toluene, respectively. The mixture, after drying, was added to an iPP matrix by extrusion. IPP/graphite melt mixing, without a masterbatch step, was performed in order to study the impact of the aPP on the dispersion. The authors observe a decrease in the crystallization of the nanocomposite with the increase of the fraction of aPP in the mixture, but especially an improvement of the dispersion of the nanofiller, with the disappearance of the large agglomerates present in the nanocomposite without masterbatch. The materials produced from the 50 wt% carbon masterbatch show excellent filler/matrix compatibility due to the aPP remaining on the surface of the fillers. In addition, it limits the reagglomeration of the carbon planes.

In the case of PP, another agent commonly used to improve nanofiller compatibility is maleic anhydride (MAH). MAH is a chemical group that can be grafted to the macromolecular chain of numerous polymers such as PP to provide new functionalities. In the case of carbon nanofiller dispersion, the polarity of the MAH grouping allows for improved dispersion, whether the blends are obtained in solution (Li *et al.* 2018), by melt mixing (Shanks and Cerezo 2012), or via a masterbatch (Li *et al.* 2017). Similar work was conducted on polyethylene by Shen *et al.* (2005). The addition of MAH to the nanocomposite elaboration opens up another elaboration possibility, allowing improved dispersion of the nanofiller through the creation of covalent bonds. One of the methods that can be used is based on the high reactivity of MAH with nitrogenous groups, like amines and amides. This reaction makes it possible, for example, to provide new functionalities to the polymer (Cui and Paul 2007), improve the mixing of a copolymer (Colbeaux *et al.* 2005) or perform matrix cross-linking (Létoffé *et al.* 2019). Hsiao *et al.* (2011) use this reactivity in a way that transforms their GO fillers into chemical cross-linking nodes, according to the reaction scheme described in Figure 1.10. A mixing without grafting operation is also performed by melt mixing for comparison. The cross-linked composites show better dispersion, better compatibility of the nanofiller with the iPP matrix, as well as an increase in the crystallinity ratio and melting temperature of the material. Several variations of this method exist in the literature, employing different amino molecules to create the matrix/filler covalent bonds (Yuan *et al.* 2014; Li *et al.* 2018). Bao *et al.* (2014), on the other hand, use octadecylamine (ODA) and demonstrate a 100% improvement in impact strength and 30% improvement in tensile strength for a PP.

1.3.4. *Comparison of development methods*

Several authors have performed the comparison between the different methods described above (Stankovich *et al.* 2006; Kalaitzidou *et al.* 2007a; Kim *et al.* 2010b; Potts *et al.* 2011; Sengupta *et al.* 2011; Gulrez *et al.* 2013). The *in situ* polymerization method appears to be the most interesting when taking the percolation threshold and final material properties into consideration. Cromer *et al.* (2015) made several iPP/GNP nanocomposites via two different methods: by *in situ* polymerization and by melt mixing. The range of materials produced by melt mixing surprisingly exhibits better dispersion, but finitely less attractive mechanical and electrical properties than its equivalent by *in situ* polymerization.

Similarly, Kalaitzidou *et al.* (2007b, 2008) have compared a conventional solution intercalation method, a filler-coated matrix method, as well as a melt-mixing method. In their case, the solution-based method, and especially the coating variant, showed better electrical properties, with a lower percolation threshold and better mechanical properties. The difference in properties is explained by a difference in morphology and dispersion, in particular by the presence of agglomerates within the melt-mixing matrix and dispersed charges in the coating-based method. The coating method has the advantage of preserving the morphology of the GNPs in the final composite, allowing extremely low percolation thresholds to be obtained. Although apparently less interesting than the other methods, the melt-mixing method remains one of the most preferred since it is the simplest to set up and does not require any solvent during the preparation. Several variants of the different methods, such as the use of masterbatch or filler-based matrix coating prior to processing, are good ways to improve the desired properties. Table 1.1 summarizes the advantages and disadvantages of the three main processing methods discussed above.

Methods	Advantages	Disadvantages
In situ polymerization	Nanofiller low percolation rate and high dispersion within the matrix. High electrical and mechanical properties. Applicable to many insoluble or thermally unstable matrices.	Requires a heavier installation than other techniques and consumes large amounts of energy. Difficult to apply on a large scale.
Solution mixing	Better dispersion state than by melt mixing, better mechanical properties.	Requires large quantities of solvent and the use of long dispersion methods (ultrasound), making industrial production difficult. Limited by the number of possible solvents.
Melt mixing	The simplest method for industrial production. Presents several variants.	Great difficulty in obtaining low percolation rates.

Table 1.1. *Advantages and drawbacks of the main preparation methods*

1.4. Influence of the nanofiller on the properties of the nanocomposite

1.4.1. *Analysis of the material morphology*

Because of the important link between the quality of the nanofiller dispersion and the mechanical and thermoelectric properties of the final nanocomposite, the analysis of the morphology of the material is a determining step of its characterization. The classical methods used for the analysis of nanofiller dispersion are transmission electron microscopy (TEM), scanning electron microscopy (SEM) and X-ray diffraction at large and small angles (WAXS and SAXS). The use of the X-ray method shows the presence of large agglomerates by different diffraction peaks, but also, in the case of graphite-derived fillers, the evolution of the distances between carbon sheets, especially for GO and EG fillers, can be studied (Kim and Macosko 2009). In the case of the dispersion of graphite fillers and its derivatives, an interesting parallel can be drawn with silicate nanofillers, which have already been widely studied in the development of nanocomposites (Paul and Robeson

2008; Ray and Okamoto 2008; Bhattacharya 2016). In both cases, the dispersion and exfoliation of the sheets are highly dependent on the preparation conditions and can result in different morphologies, as illustrated in Figure 1.11.

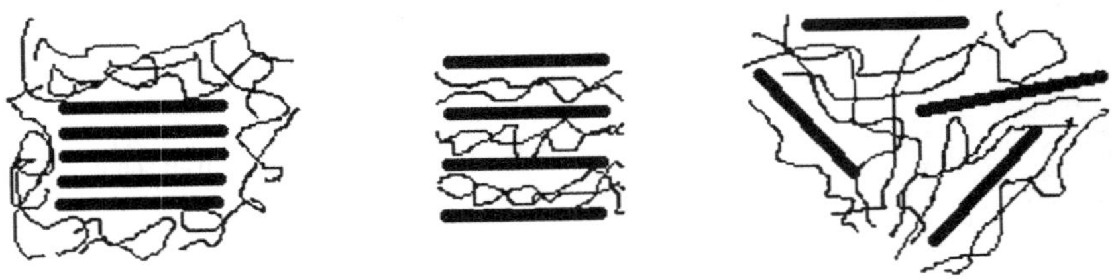

a) b) c)

Figure 1.11. *Representations of the different morphologies of nanocomposites encountered in the case of silicate nanofillers that can be applied to graphite. Phases: a) separated, b) intercalated and c) exfoliated (Bhattacharya 2016)*

For example, Huang *et al.* (2010) employ both SEM and TEM imaging techniques to characterize the morphology of their PP/GO composites produced by *in situ* polymerization. Their preparation method based on the surface grafting of their GOs with Ziegler–Natta catalysts enables the surface polymerization of carbon planes promoting their exfoliation. Using these methods, it is possible to highlight the presence of nanofillers and characterize their states of dispersion, but also estimate the number of planes composing the GO fillers after preparation, as well as appreciate the quality of the adhesion of the matrix to the filler. It should be noted, however, that some morphologies are difficult to observe, especially in the case of particularly exfoliated fillers, namely in the form of perfectly isolated sheets in the matrix, either by TEM or X-ray (Paul and Robeson 2008; Kim *et al.* 2010b). In addition, the cryomicrotomy cutting necessary for sampling can prove complicated depending on the matrix employed.

1.4.2. *Influence of nanofillers on semi-crystalline microstructures*

In the case of a semicrystalline matrix, the carbon filler can impact the crystallinity rate, the thicknesses of crystalline lamella and/or the crystallization speed. In the case of PP, carbon planes behave as nucleating agents, but limit macromolecular mobility and thus crystal lamella growth (Kalaitzidou *et al.* 2008; Yuan *et al.* 2014; He *et al.* 2018). Filler compatibility with the matrix, as well as processing and dispersion conditions play a very important role. Therefore, in their work comparing *in situ* polymerization and melt-mixing processing methods, Cromer *et al.* (2015) show that in both situations an increase in the melting temperature of their iPPs results in a higher crystallinity rate as a function of the increasing mass fraction of carbon particles. However, this result is twice as large for materials obtained by melt mixing. In contrast, Bafana *et al.* (2017) show a decrease in the crystallinity rate and only a very small variation in the melting temperature for their materials produced by a solution-mixing method. On the other hand, Hsiao *et al.* (2011), in their work on nanofiller grafting, show a strong difference between PP/TRGO materials produced by simply dispersing their nanofillers, and PP/PP-g-TRGO materials, in which the matrix and the nanofiller exhibit covalent bonds and thus better compatibility. This difference in behavior highlights the importance of the compatibility between the matrix and the filler in the production of nanocomposites.

The crystallization rate of a matrix is also strongly impacted by the insertion of fillers in a semi-crystalline matrix. In the case of PP, several authors have studied the effect of graphite fillers and its derivatives on the crystallization rate, as well as on the final morphology of the material depending on the amount, but also the size of the nanofiller (Kalaitzidou *et al.* 2008; Zhao *et al.* 2014; Jun *et al.* 2018a). Zhao *et al.* (2014) observe that the presence of graphite, even at low concentration, allows surface, but especially edge nucleation of carbon sheets at temperatures higher than pure materials. The preferential localization at the edge of the sheets can be explained by the presence of a greater number of defects in the structure of the filler, making a better nucleation possible. The authors also observe a decrease in the size of the crystalline structures due to their greater number and the space limited by the presence of graphite. These results are confirmed in the work of Kalaitzidou *et al.* (2008) on the crystallization of a PP charged with different graphite fillers. The crystallization rate increases significantly with the addition of graphite: it can triple between a virgin PP and a PP/GNP at 0.01 wt%, and sixfold for concentrations up to 1 wt%. This increase of the crystallization rate, but also of the number of spherulites developing around the nanofillers, implies a decrease in their final sizes.

The effect of the size of the graphite filler in terms of diameter is also studied by Kalaitzidou *et al.* (2008). Their optical analyses show that at equivalent loading rates, the materials with the largest loads crystallize as larger spherulites, while exhibiting close crystallinity rates. The X-ray diffraction analysis shows a significant difference between the materials containing a filler of 1 μm and 15 μm diameter. While increasing the concentration of both fillers impacts the α-phase of iPP in the same way – an increase in the intensity of the diffraction peaks of the plane family (00l) is observed – a difference appears for the *β-phase* confirmed by Jun *et al.* (2018b). The 1 μm diameter fillers at low concentrations (0.01 wt%) enable the crystallization of the material under its *β* phase, but not at higher concentrations (1 wt%). On the contrary, the addition of the 15 μm diameter filler systematically induces a *β-phase* even at higher concentrations (e.g. 10 wt%). This difference comes from the large number of nucleation points created by the 1 μm diameter fillers, which by rapidly occupying the available space do not allow the formation of *β-spherules*. The appearance of the *β* phase is also indicated by other studies (Shanks and Cerezo 2012; Ruy and Shanmugharaj 2014; Ahmad *et al.* 2017; He *et al.* 2018; Rosehr *et al.* 2018). In general, the nucleating effect is higher for smaller fillers because they provide a larger specific surface area of contact with the matrix and limit the movement of polymer chains less (Vallès *et al.* 2014).

1.4.3. *Influence of nanofillers on mechanical properties*

One of the main reasons for adding a carbon or another nanofiller to a polymer matrix is to improve its mechanical properties. Therefore, the effect of carbon nanofillers on the mechanical properties of polymer matrices is studied in many formulations, and as in the case of the influence on the microstructure, the effects vary greatly depending on the chosen configuration. For example, exfoliated graphene fillers appear to be more suitable for mechanical reinforcements than black or carbon nanotube fillers (Kim *et al.* 2010a; Bhattacharya 2014). Similarly, the compatibility between filler and matrix can greatly influence the mechanical properties of the materials. Polar polymers such as PVA, PMMA and PAN, but also epoxy matrices, are known for large increases in their Young's moduli and glass transition temperatures with the addition of carbon nanofillers, especially GO, due to the non-covalent bonds that can be created between the matrix and the chemical groups present on the surface of the carbon particles (Ramanathan *et al.* 2008; Rafiee *et al.* 2009; Ren *et al.* 2018). In general, carbon nanofillers induce an increase in the Young modulus and tensile strength of the material, parallel to the rate under load. Thus, it is generally accepted that the nanofiller limits the movement of the macromolecular chains of the material, leading to its stiffening. This evolution is generally achieved to the detriment of the elongation at break of the material and depends on the adhesion between filler and matrix.

These trends are observed in the case of PP-based nanocomposites. For example, Jun *et al.* (2018a) show that the addition of large-diameter GNPs (>150 μm) at 1.3 wt% allows the increase of the Young modulus of their PP matrices from 514 to 540 MPa, but strongly decreases its elongation at break, ranging from 908% to only

17%. Some authors also note a loss of ductility of the nanocomposite (Song *et al.* 2011; Yuan *et al.* 2014; Rosehr *et al.* 2018). This decrease can be explained by the tendency of the fillers to agglomerate, making decohesion with the matrix easier. Zhao *et al.* (2010) compare this condition to filler "mechanical percolation". Improved filler dispersion and exfoliation yields better mechanical properties. Cromer *et al.* (2015), in their comparisons of *in situ* polymerization and melt-mixing methods, note that at equivalent filler rates a better ductility of their matrices is produced by the first method, for which a better dispersion state was observed. Similarly, Kalaitzidou *et al.* (2007b, 2008) observe an improvement of the mechanical properties of their composites via their method of elaboration by coating, also resulting from better filler dispersion.

In addition to the dispersion quality, the size of the fillers has an influence on the final mechanical properties of the nanocomposite. As previously stated, it influences PP crystallization, in particular by promoting the development of the *β-crystalline* phase which, although the latter presents a better impact strength than the *α-phase*, shows lower mechanical tensile strength. More generally, Young *et al.* (2018) explain that the stress transfer between the matrix and the nanofiller and the Young modulus of the nanocomposites are globally independent of the intrinsic properties of the nanofiller, but mainly depend on their form factor and compatibility with the matrix. Jun *et al.* (2018b) thus compare four types of GNPs with different diameters: 2 μm (C300), 5 μm (M5), 25 μm (M25) and 150 μm (H100), for mass concentrations ranging from 1 to 10 wt%. The results of their mechanical analyses are presented in Figure 1.12. The range of nanocomposites filled with H100 GNPs shows an increase in Young's modulus with the carbon concentration, but a low tensile strength, while the other three ranges show an improvement of both parameters, but all at the expense of the elongation at break. This evolution in H100 materials is explained by the large size of the carbon sheets that fold and form large agglomerates. On the contrary, the materials based on the smaller filler sizes show an improvement in both their Young's moduli and their tensile strength, with a maximum reached for the C300 filler, the smallest in this study. Therefore, decreasing the size of the filler makes it possible, at equivalent loading rates, to increase the contact area and thus achieve a better stress transfer between the matrix and the filler (Kuvardina *et al.* 2013; Liang *et al.* 2016).

Finally, the last parameter that can influence the mechanical properties is the compatibility of the matrix with the surface of the filler. A compatibility agent is then used as an intermediary to yield a better mechanical filler transfer.

Rosehr *et al.* (2018) employ atactic PP chains as compatibilizing agents on the surface of GNPs when performing a masterbatch. After mixing within an iPP matrix, the resulting materials exhibit tensile strength equivalent to a composite produced without a masterbatch step, but show higher elongation at break, as shown in Figure 1.13. It is interesting to note that the elongation at break is all the higher when the aPP mass fraction is large. The presence of aPP chains on the surface of the carbon ensures a better transfer of the stress and limits the phenomenon of filler/matrix decohesion within the nanocomposite.

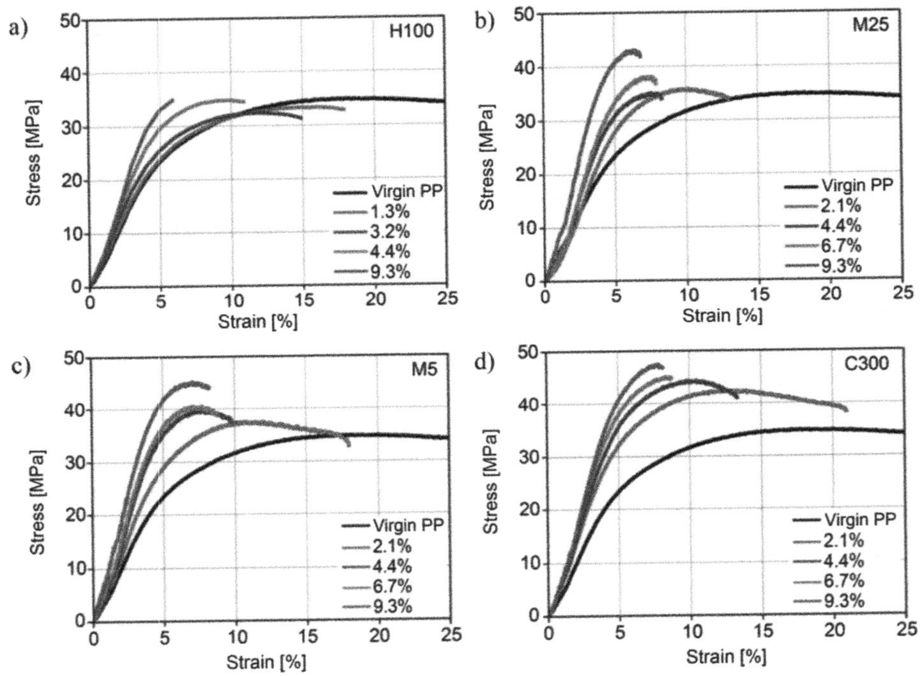

Figure 1.12. *Nominal behavior laws of different PP/GNP composites with the fillers: a) H100, b) M25, c) M5, and d) C300, for a volume concentration ranging from 1.3 to 9.3% (Jun* et al. *2018b). For a color version of this figure, see www.iste.co.uk/bai/nanocomposites.zip*

Other authors use grafting of different functional chemical groups such as maleic anhydride onto the macromolecular chains. The resulting chemical adhesion induces a net improvement of the dispersion of the filler within its matrix, but also allows the significant improvement of its mechanical properties (Colbeaux *et al.* 2005; Li *et al.* 2017). The main drawback of these methods originates from the impact of these groups and their grafting operations on the properties of the pristine matrix. Other types of grafting are listed in the literature, and those inducing covalent bonds with the filler provide significant improvement in mechanical properties (Hsiao *et al.* 2011; Ryu and Shanmugharaj 2014).

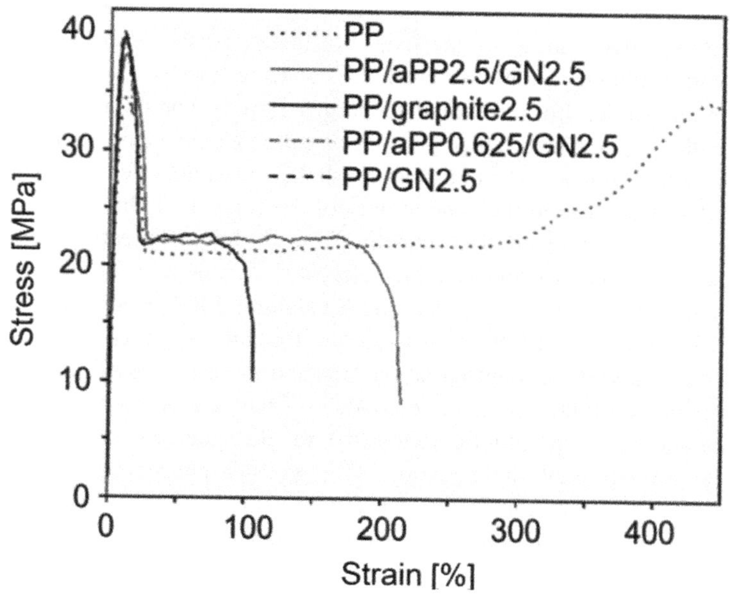

Figure 1.13. *Nominal mechanical behavior of a PP and of different composites (Rosehr* et al. *2018)*

1.4.4. *Influence of nanofillers on electrical properties*

Regardless of their shapes, the choice of carbon nanofillers in the development of a nanocomposite is mainly due to their excellent electrical properties compared to other fillers (Bhattacharya 2016). However, in order to avoid having to add them in too large quantities, which could adversely affect other properties of the final material, the best method to increase conductivity is to improve dispersion by reducing the agglomeration phenomenon. This strong impact of dispersion is explained by the phenomenon of percolation of the charge within the matrix by the creation of a necessary and essential conductive lattice (Kirkpatrick 1973; Lux 1993).

1.4.4.1. *The role of percolation*

During the preparation of a nanocomposite, three states are possible, according to the rate of charge and the dispersion of the latter. At a low loading rate, the fillers are separated by the insulating matrix, the conductivity of the material is then close to that of the initial matrix. As the load rate increases, the filler-to-filler contact becomes more frequent, until a threshold concentration is reached, called the percolation threshold, from which the charges form a continuous lattice in the material, called a percolated lattice. The electrical conductivity of the material then presents a strong increase. Beyond this threshold, the conductivity of the material stabilizes at values close to the conductivity of the filler. In addition to this contact conductivity, the phenomenon of "tunneling" occurs for sufficiently close charges, for distances less than 10 nm, enabling electrical conductivity before the contact percolated lattice is formed (Al-Saleh and Sundararaj 2009; Stauffert *et al.* 2009). Figure 1.14 shows the different possible configurations. In case of poor filler distribution and dispersion (configuration (a)), and in case of good distribution but poor dispersion (configuration (b)), no conductive lattice is formed. In case of good distribution and dispersion (configuration (d)), the distance between the charges can prevent the formation of the network. At low filler concentration, the optimal configuration is therefore configuration (c), with good dispersion, but a distribution of fillers enabling the formation of the continuous network.

Many parameters are involved in the percolation phenomenon, whether it is the polymer matrix used, its compatibility with the filler, the preparation method, or the form factor of the filler. In their review on the production of polymer/graphene nanocomposites, Potts *et al.* (2011) compare the lowest electrical percolation thresholds obtained with several polymer/filler configurations. Their observations are summarized in Table 1.2. Significant variations in the percolation threshold are observed, down to values below 0.2 wt% for an oxidized polyethylene/graphene composite, as seen for the solution-based preparation by Pang *et al.* (2010).

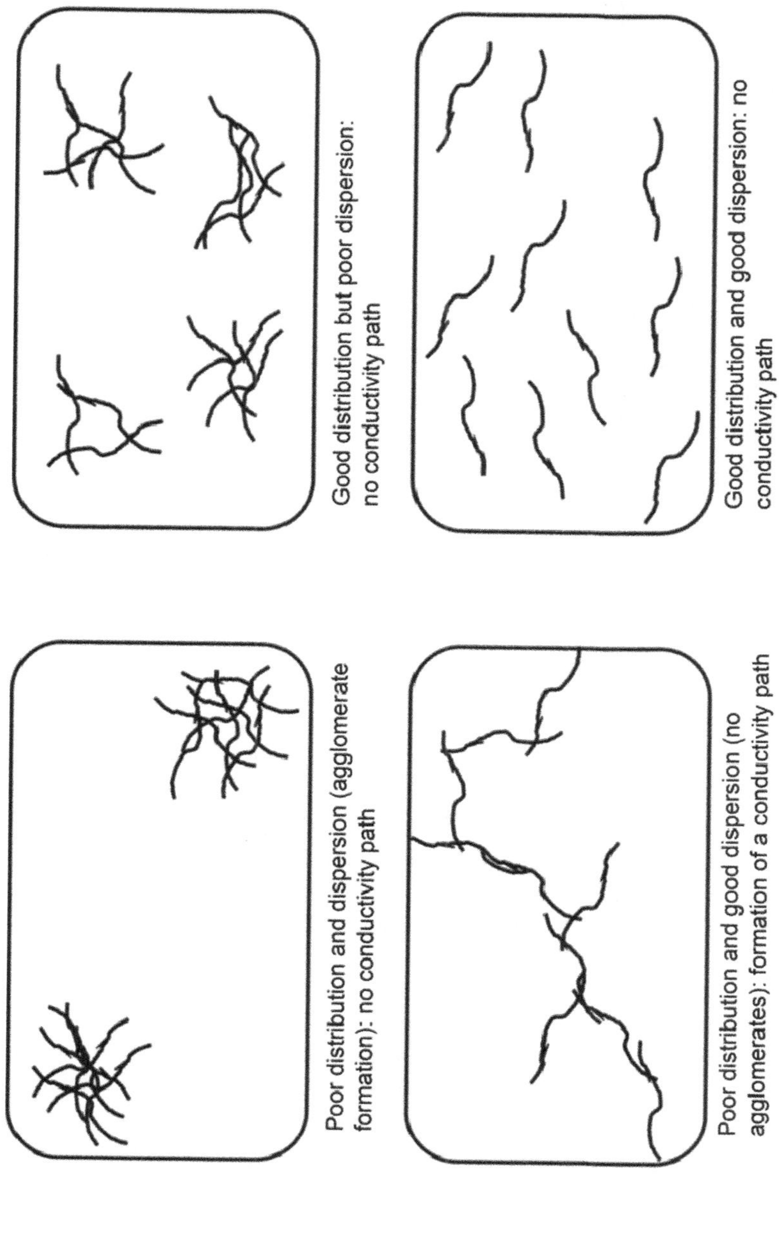

Figure 1.14. *Schematic representation of the possible distributions and dispersions of fillers within a polymer matrix (Al-Saleh and Sudararaj 2009)*

Polymer matrix	Nanofiller	Percolation threshold (wt%)	Reference
Epoxy	Fonct. EG	1.0	(Miller *et al.* 2010)
Nylon-6	GO	0.5	(Du *et al.* 2010)
Polyaniline (*doped*)	GNP	0.7	(Wu *et al.* 2010b)
Polycarbonate	TEGO	0.3	(Khanam *et al.* 2016)
Polyethylene	RG-O	0.15	(Pang *et al.* 2010)
Polyethylene terephthalate	TEGO	1.0	(Zhang *et al.* 2010)
Polymethyl methacrylate	GNP	0.7	(Chen *et al.* 2003a)
Polypropylene	GNP	0.7	(Kalaitzidou *et al.* 2007b)
Polystyrene	Fonct. GO	0.2	(Stankovich *et al.* 2006)
Polyvinyl alcohol	RG-O	0.5	(Salavagione *et al.* 2009)
Polyvinyl chloride	GNP	1.4	(Vadukumpully *et al.* 2011)
Polyvinylidene fluoride	TEGO	2.0	(Ansari *et al.* 2009)
Polyurethane	TEGO	0.6	(Kim *et al.* 2010b)

Table 1.2. *Comparison of different percolation rates obtained as a function of the matrix/filler configurations (Potts* et al. *2011)*

1.4.4.2. *Evolution of electrical properties*

As in the case of the influence on the microstructure and on the mechanical properties, many parameters influence the electrical properties of the nanocomposite, such as the size of the charge, its dispersion, the compatibility with the matrix, but also in the case of semi-crystalline materials, the crystallinity of the material. It is possible, depending on the configuration, to reach electrical conductivities of the order of $S \cdot cm^{-1}$ (Kalaitzidou *et al.* 2007b; Jun *et al.* 2018a).

He *et al.* (2018) show that the use of ultrasound during the preparation of their materials allows an increase in the conductivity of the materials, resulting in a better dispersion. This difference in percolation, according to the quality of the dispersion, is particularly visible in the work by Kalaitzidou *et al.* (2008), who compare the levels of dispersion obtained according to the methods of elaboration in solution and by melt mixing. The latter shows a percolation threshold above 5 vol%, while the solution method gives a percolation between 0 and 3 vol% (Figure 1.15).

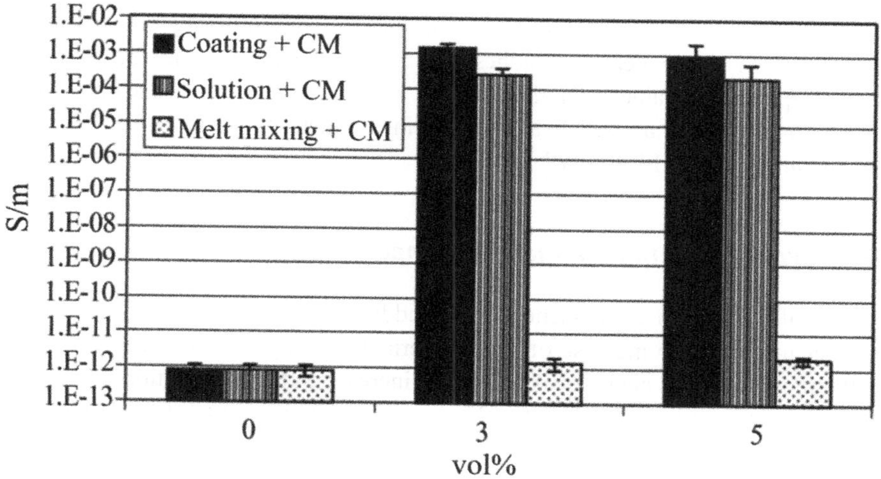

Figure 1.15. *Evolution of electrical conductivity in PP/GNP nanocomposites according to the preparation method used (Kalaitzidou* et al. *2008)*

Jun *et al.* (2018a) show a difference in the electrical properties of their PP/GNP materials depending on the size of the introduced fillers. The first observation made by the authors is the lack of change in electrical conductivity with the addition of their smallest filler (2 μm diameter). The larger the filler diameter, the lower the percolation threshold, which is in agreement with previous results on the filler size/percolation threshold relation proposed by Wu *et al.* (2013) and Jun *et al.* (2015). It is interesting to note that this evolution is to the advantage of large diameter fillers and is in contrast to the previously observed evolutions related to microstructure and mechanical properties, both of which are more favorable to small-sized fillers. A trade-off must then be found between electrical and mechanical properties.

Kalaitzidou *et al.* (2008) show that the cooling rate of the material during its preparation strongly impacts the percolation threshold of the nanocomposite. Indeed, the percolation threshold is four to five times lower for materials cooled slowly. This difference is explained at the microstructural level. During fast cooling, a large number of small spherulites form on the surface of the carbon films and block the formation of the electrical percolation lattice, whereas slow cooling allows for the formation of large spherulites favorable to percolation.

A similar finding can be made for nanocomposites utilizing compatibilizing agents that allow covalent-type adhesion between the matrix and the filler. The better dispersion obtained via this method is responsible for a decrease in the percolation threshold, as well as an increase in conductivity (Hsiao *et al.* 2011; Ryu and Shanmugharaj 2014; Li *et al.* 2017).

1.4.5. *Evolution of the thermal resistance*

One of the known consequences of the addition of a carbon nanofiller within a polymer matrix is the increase of matrix thermal stability. This can be observed in air thermogravimetric analysis (ATG) by an increase in the degradation temperature of the material with the filler ratio. For example, Bafana *et al.* (2017) obtain an increase in the degradation temperature of their iPP with the addition of graphene (Figure 1.16).

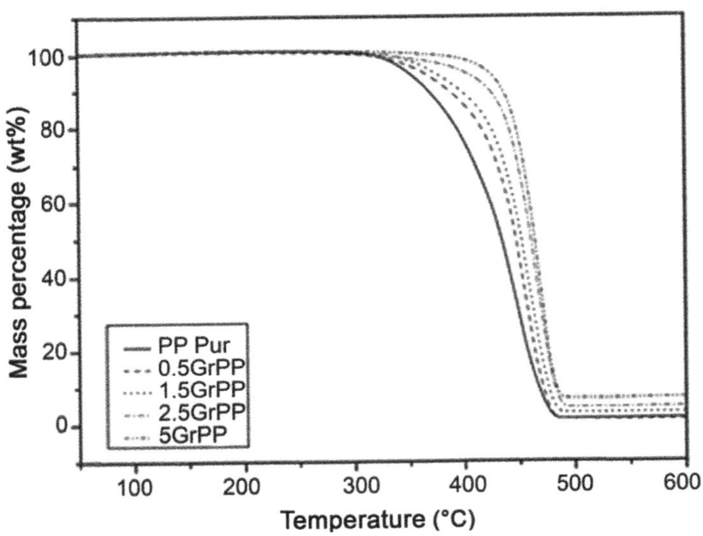

Figure 1.16. *Evolution of mass losses measured by ATG under air of pure iPP and its nanocomposites loaded with different graphene filler rates (Bafana* et al. *2017). For a color version of this figure, see www.iste.co.uk/bai/nanocomposites.zip*

This increase in material degradation temperature is not linear with increasing loading rate, but increases significantly from low values and then reaches an asymptote (Song *et al.* 2011; Kuvardina *et al.* 2013; Zhao *et al.* 2014; Ahmad *et al.* 2017; Ren *et al.* 2018). This peculiarity can be explained by the role of nanofillers in the permeability to gases, more specifically to the oxygen of the nanocomposite. Indeed, the graphene sheets, if devoid of defects, present an impermeability to gaseous molecules and thus make it possible to limit their diffusion within, and through, the polymer matrix (Bunch *et al.* 2008). The resulting nanocomposites then become good candidates as protective materials against corrosion, in the form of an impermeable film (Yang *et al.* 2013). At equivalent filler ratio, the effect is more important for small diameter fillers (Figure 1.17) due to the increase in the interaction surface between the filler and the matrix (Vadukumpully *et al.* 2011; Beuguel *et al.* 2018). Improved compatibility between matrix and filler implies better thermal energy diffusion in the matrix and thus also improves the thermal resistance of the composite. On the other hand, Hsiao *et al.* (2011) find that the addition of 0.1wt% of a TRGO filler induces an increase in the thermal degradation temperature of an iPP. This property is only promoted if the filler is grafted to the matrix.

Finally, we note that the large difference between the degradation temperatures of the polymer and the nanofiller enables, after preparation, a verification of the mass percentage remaining after complete degradation of the polymer matrix.

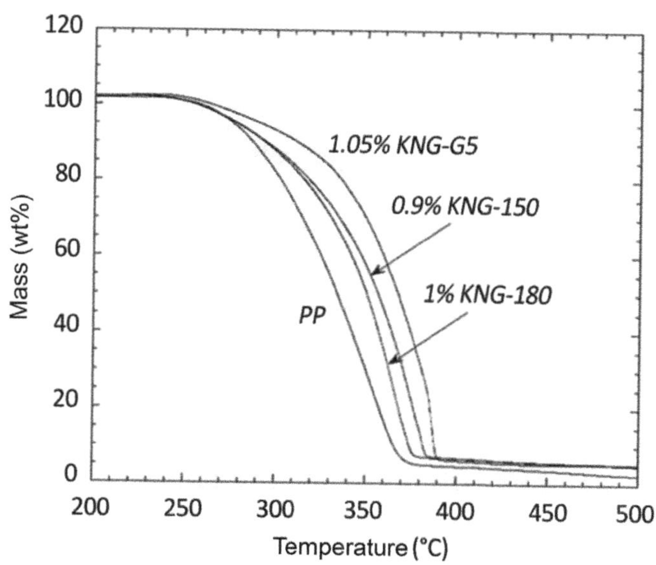

Figure 1.17. *ATG analyses under air of different PP/GNP composites for GNPs of different sizes (increasing order of size: G5, KNG 150, KNG 180) (Beuguel* et al. *2018)*

1.5. References

Afanasov, I.M., Shornikova, O.N., Kirilenko, D.A. (2010). Graphite structural transformations during intercalation by HNO_3 and exfoliation. *Carbon*, 48(6) 1862–1865.

Ahmad, S.R., Xue, C., Young, R.J. (2017). The mechanisms of reinforcement of polypropylene by graphene nanoplatelets. *Materials Science and Engineering: B*, 216, 2–9.

Al-Saleh, M.H. and Sundararaj, U. (2009). A review of vapor grown carbon nanofiber/polymer conductive composites. *Carbon*, 47, 2–22.

Anders, A. (2009). *Cathodic Arcs: From Fractal Spots to Energetic Condensation*. Springer, New York.

Ansari, S. and Giannelis, E.P. (2009). Functionalized graphene sheet-poly(vinylidene fluoride) conductive nanocomposites. *Journal of Polymer Science Part B: Polymer Physics*, 47(9), 888–897.

Bae, S., Kim, H., Lee, Y. (2010). Roll-to-roll production of 30-inch graphene films for transparent electrodes. *Nature Nanotechnology*, 5(8), 574–578.

Bafana, A.P., Yan, X., Wei, X., Patel, M., Guo, Z., Wei, S., Wujcik, E.K. (2017). Poly-propylene nanocomposites reinforced with low weight percent graphene nanoplatelets. *Composites: Part B*, 109, 101–107.

Bagri, C., Matteri, C., Acik, M. (2010). Structural evolution during the reduction of chemically derived graphene oxide. *Nature Chemistry*, 2(7), 581–587.

Banhart, F., Kotakoski, J., Krasheninnikov, A.V. (2011). Structural defects in graphene. *ACS Nano*, 5(1), 26–41.

Bao, R.Y., Cao, J., Liu, Z.Y., Yang, W., Xie, B.X., Yang, M.B. (2014). Towards balanced strength and toughness improvement of isotactic polypropylene nanocomposites by surface functionalized graphene oxide. *Journal Materials Chemistry: A*, 2(9), 3190–3199.

Becerril, H.A., Mao, J., Liu, Z. (2008). Evaluation of solution-processed reduced graphene oxide films as transparent conductors. *ACS Nano*, 2(3), 463–470.

Berry, V. (2013). Impermeability of graphene and its applications. *Carbon*, 62, 1–10.

Beuguel, Q., Mija, A., Vergnes, B., Peuvrel-Disdier, E. (2018). Structural, thermal, rheological and mechanical properties of polypropylene/graphene nanoplatelets composites: Effect of particle size and melt mixing conditions. *Polymer Engineering and Science*, 58(11), 1937–1944.

Bhattacharya, M. (2016). Polymer nanocomposites: A comparison between carbon nanotubes, graphene, and clay as nanofillers. *Materials*, 9(4), 262.

Bianco, A., Cheng, H.M., Enoki, T., Gogotsi, Y., Hurt, R.H., Koratkar, N., Kyotani, T., Monthioux, M., Park, C.R., Tascon, J.M.D., Zhang, J. (2013). All in the graphene family – A recommended nomenclature for two-dimensional carbon materials. *Carbon*, 65, 1–6.

Boehm, H.P., Clauss, A., Fischer, G. (1962). Surface properties of extremely thin graphite lamellae. *Proceedings of the Fifth Conference on Carbon*, 73–80.

Bolotin, K.I., Sikes, K.J., Jiang, Z., Klima, M., Fudenberg, G., Hone, J. (2008). Ultra-high electron mobility in suspended graphene. *Solid State Communications*, 146(9), 351–355.

Botas, C., Alvarez, P., Blanco, C., Santamaria, R., Granda, M., Ares, P., Rodriguez-Reinoso, F., Menendez, R. (2012). The effect of the parent graphite on the structure of graphene oxide. *Carbon*, 50(1), 275–282.

Botas, C., Alvarez, P., Blanco, P., Granda, M., Blanco, C., Santamaria, R., Romasanta, L.J., Verdejo, R., Lopez-Manchado, M., Menendez, R. (2013). Graphene materials with different structures prepared from the same graphite by the Hummers and Brodie methods. *Carbon*, 65, 156–164.

Brault, P., Caillard, A., Baranton, S., Mougenot, M., Cuynet, S., Coutanceau, C. (2013). One-step synthesis and chemical characterization of Pt–C nanowire composites by plasma sputtering. *ChemSusChem*, 6(7), 1168–1171.

Bruggeman, P.J., Kushner, M.J., Locke, B.R., Gardeniers, J.G.E., Graham, W.G., Graves, D.B., Hofman-Caris, R.C.H.M., Maric, D., Reid, J.P., Ceriani, E. (2016). Plasma–liquid interactions: A review and roadmap. *Plasma Sources Science and Technology*, 25(5), 053002.

Bunch, J.S., Verbridge, S.S., Alden, J.S., Van der Zande, A.M., Parpia, J.M., Craighead, H.G., McEuen, P.L. (2008). Impermeable atomic membranes from graphene sheets. *Nano Letters*, 8(8), 2458–2462.

Canales, J., Fernadez, M., Pena, J.J., Munoz, M.E., Santamaria, A. (2015). Rheological methods to investigate graphene/amorphous polyamide nanocomposites: Aspect ratio, processing, and crystallization. *Polymer Engineering and Science*, 55(5), 1142–1151.

Catheline, A., Vallés, C., Drummond, C., Ortolani, L., Morandi, V., Marcaccio, M., Iurlo, M., Paolucci, F., Pénicaud, A. (2011). Graphene solutions. *Chemical Communications*, 47(19), 5470–5472.

Catheline, A., Ortolani, L., Morandi, V., Melle-Franco, M., Drummond, C., Zakri, C., Pénicaud, A. (2012). Solutions of fully exfoliated individual graphene sheets in low boiling point solvents. *Soft Matter*, 8(30), 7882–7887.

Chen, G., Weng, W., Wu, D., Wu, C. (2003a). PMMA/graphite nanosheets composite and its conducting properties. *European Polymer Journal*, 39(12), 2329–2335.

Chen, G., Wu, C., Weng, W., Wu, D., Yan, W. (2003b). Preparation of polystyrene/ graphite nanosheet composite. *Polymer*, 44(6), 1781–1784.

Colbeaux, A., Fenouillot, F., Gerard, J.F., Taha, M., Wautier, H. (2005). Diamine coupling of maleic anhydride modified polyethylene. *Polymer International*, 54(4), 692–697.

Cromer, B.M., Scheel, S., Luinstra, G.A., Coughlin, E.B., Lesser, A.J. (2015). *In situ* polymerization of isotactic polypropylene-nanographite nanocomposites. *Polymer*, 80, 275–281.

Cui, L. and Paul, D.R. (2007). Evaluation of amine functionalized polypropylenes as compatibilizers for polypropylene nanocomposites. *Polymer*, 48(6), 1632–1640.

Dey, A., Chroneos, A., Braithwaite, N.S.J., Gandhiraman, R.P., Krishnamurthy, S. (2016). Plasma engineering of graphene. *Applied Physics Reviews*, 3(2), 021301.

Dhar, S., Barman, A.R., Ni, G.X. (2011). A new route to graphene layers by selective laser ablation. *AIP Advances*, 1(2), 022109.

Dreyer, D.R., Park, S.J., Bielawski, C.W. (2010). The chemistry of graphene oxide. *Chemical Society Reviews*, 39(1), 228–240.

Du, N., Zhao, C.Y., Chen, Q., Wu, G., Lu, R. (2010). Preparation and characterization of nylon 6/graphite composite. *Materials Chemistry and Physics*, 120(1), 167–171.

Fang, M., Wang, K., Lu, H., Yang, Y., Nutt, S. (2009). Covalent polymer functionalization of graphene nanosheets and mechanical properties of composites. *Journal of Materials Chemistry*, 19(38), 7098–7105.

Fernandez-Merino, M.J., Guardia, L., Paredes J.I. (2010). Vitamin C is an ideal substitute for hydrazine in the reduction of graphene oxide suspensions. *Journal of Physical Chemistry C*, 114(14), 6426–6432.

Ferrari, A.C., Bonaccorso, F., Fal'Ko, V.I., Novoselov, K.S., Roche, S., Boggild, P., Garrido, J.A. (2015). Science and technology roadmap for graphene, related two-dimensional crystals, and hybrid systems. *Nanoscale*, 7(11), 4598–4810.

Figuereido, J.L., Pereira, M.F.R., Freitas, M.M.A. (1999). Modification of the surface chemistry of activated carbons. *Carbon*, 37(9), 1379–1389.

Fim, F.C., Guterres, J.M., Basso, N.R.S., Galland, G.B. (2010). Polyethylene/graphite nanocomposites obtained by *in situ* polymerization. *Journal of Polymer Science: Part A: Polymer Chemistry*, 48, 692–698.

Furdin, G. (1998). Exfoliation process and elaboration of new carbonaceous materials. *Fuel*, 77(6), 479–485.

Furdin, G., Marêché, J.F., Hérold, A. (1991). Graphite micronique plat, procédé pour sa préparation et ses applications. Brevet français no. 91-12663.

Gao, W., Alemany, L.B., Ci, L. (2009). New insights into the structure of graphite oxide. *Nature Chemistry*, 1(5), 403–408.

Giorcelli, M., Guastella, S., Mandracci, P., Liang, Y., Li, X., Tagliaferro, A. (2018). Carbon fibre functionalization by plasma treatment for adhesion enhancement on polymers. *AIP Conference Proceedings*, 1981, 020142.

Gomez-Navarro, C., Weitz, R.T., Bittner, A.M. (2007). Electronic transport properties of individual chemically reduced oxide graphene sheets. *Nano Letters*, 7(11), 3499–3503.

Green, A.A. and Hensen, M.C. (2009). Solution phase production of graphene with controlled thickness via density differentiation. *Nano Letters*, 9(12), 4031–4036.

Gulrez, S.K.H., Mohsin, M.E.A., Shaikh, H., Anis, A., Pulose, A.M., Yadav, M.K., Qua, A.H.P., Al-Zahrani, S.M. (2013). A review on electrically conductive polypropylene and polyethylene. *Polymer Composites*, 35, 900–914.

Hamdan, A., Marinov, I., Rousseau, A., Belmonte, T. (2013a). Time-resolved imaging of nanosecond-pulsed micro-discharges in heptane. *Journal of Physics D: Applied Physics*, 47, 055203.

Hamdan, A., Noel, C., Kosior, F., Henrion, G., Belmonte, T. (2013b). Dynamics of bubbles created by plasma in heptane for micro-gap conditions. *The Journal of the Acoustical Society of America*, 134, 991–1000.

He, S., Zhang, J., Xiao, X., Hong, X. (2018). Effects of ultrasound vibration on the structure and properties of polypropylene/graphene nanoplatelets composites. *Polymer Engineering and Science*, 58(3), 377–386.

Hernandez, Y., Nicolosi, V., Lotya, M. (2008). High-yield production of graphene via liquid-phase exfoliation of graphite. *Nature Nanotechnology*, 3(9), 563–568.

Hsiao, M.C., Liao, S.H., Lin, Y.F., Wang, C.A., Pu, N.W., Tsai, H.M., Ma, C.C. (2011). Preparation and characterization of polypropylene-graft-thermally reduced graphite oxide with an improved compatibility with polypropylene-based nanocomposite. *Nanoscale*, 3(4), 1516–1522.

Huang, Y.J., Qin, Y., Zhou, Y., Niu, H., Yu, Z.Z., Dong, J.Y. (2010). Polypropylene/graphene oxide nanocomposites prepared by *in situ* Ziegler–Natta polymerization. *Chemistry of Materials*, 22(13), 4096–4102.

Hummers, W.S. and Offeman, R.E. (1958). Preparation of graphitic oxide. *Journal of the American Chemical Society*, 208(1937), 1339–1339.

Inagaki, M. and Tashiro, R. (2004). Exfoliation process of graphite via intercalation compounds with sulfuric acid. *Journal of Physics and Chemistry of Solids*, 65(2–3), 133–137.

Jang, B.Z. and Zhamu, A. (2008). Processing of nanographene platelets (NGPs) and NGP nanocomposites: A review. *Journal of Material Science*, 43(15), 5092–5101.

Jun, Y.S., Sy, S., Ahn, W., Zarrin, H., Rasen, L., Tjandra, R., Amoli, B.M., Zhao, B., Chiu, G., Yu, A. (2015). Highly conductive interconnected graphene foam based polymer composite. *Carbon*, 95, 653–658.

Jun, Y.S., Um, J.G., Jiang, G., Lui, G., Yu, A. (2018a). Ultra-large sized graphene nanoplatelets (GnPs) incorporated polypropylene (PP)/GnPs composites engineered by melt compounding and its thermal, mechanical, and electrical properties. *Composites: Part B*, 133(15), 218–225.

Jun, Y.S., Um, J.G., Jiang, G., Yu, A. (2018b). A study on the effects of graphene nanoplatelets (GnPs) sheet sizes from a few to hundred microns on the thermal, mechanical, and electrical properties of polypropylene (PP)/GnPs composites. *eXPRESS Polymer Letters*, 12(10), 885–897.

Kalaitzidou, K., Fukushima, H., Drzal, L.T. (2007a). Multifunctional polypropylene composites produced by incorporation of exfoliated graphite nanoplatelets. *Carbon*, 45(7), 1446–1452.

Kalaitzidou, K., Fukushima, H., Drzal, L.T. (2007b). A new compounding method for exfoliated graphite–polypropylene nanocomposites with enhanced flexural properties and lower percolation threshold. *Composites Science and Technology*, 67(10), 2045–2051.

Kalaitzidou, K., Fukushima, H., Askeland, P., Drzal, L.T. (2008). The nucleating effect of exfoliated graphite nanoplatelets and their influence on the crystal structure and electrical conductivity of polypropylene nanocomposites. *Journal of Materials Science*, 43(8), 2895–2907.

Kernina, A., Wana, K., Liu, Y., Shi, X., Kong, J., Bilotti, E., Peijs, T., Zhang, H. (2019). The effect of graphene network formation on the electrical, mechanical, and multi-functional properties of graphene/epoxy nanocomposites. *Composites Science and Technology*, 169, 224–231.

Khan, U., O'Neill, A., Lotya, M. (2008). High-concentration solvent exfoliation of graphene. *Small*, 6(7), 864–871.

Khanam, P.N., Al-Maadeed, M.A., Ouederni, M., Harkin-Jones, E., Mayoral, B., Hamilton, A., Sun, D. (2016). Melt processing and properties of linear low density polyethylenegraphene nanoplatelet composites. *Vacuum*, 130, 63–71.

Kim, H. and Macosko, C.W. (2009). Processing-property relationships of polycarbonate/ graphene composites. *Polymer*, 50(15), 3797–3809.

Kim, J. and Suh, J.S. (2014). Size-controllable and low-cost fabrication of graphene quantum dots using thermal plasma jet. *ACS Nano*, 8(5), 4190–4196.

Kim, H., Abdala, A.A., Macosko, C.W. (2010a). Graphene/polymer nanocomposites. *Macromolecules*, 43(16), 6515–6530.

Kim, H., Miura, Y., Macosko, C.W. (2010b). Graphene/polyurethane nanocomposites for improved gas barrier and electrical conductivity. *Chemistry of Materials*, 22(11), 3441–3450.

Kirkpatrick, S. (1973). Percolation and conduction. *Reviews of Modern Physics*, 45(4), 574–588.

Kuvardina, E.V., Novokshonova, L.A., Lomakin, S.M., Timan, S.A., Tchmutin, I.A. (2013). Effect of the graphite nanoplatelet size on the mechanical, thermal, and electrical properties of polypropylene/exfoliated graphite nanocomposites. *Journal of Applied Polymer Science*, 128(3), 1414–1424.

Lee, C., Wei, X., Kysar, J.W., Hone, J. (2008). Measurement of the elastic properties and intrinsic strength of monolayer graphene. *Science*, 321(5887), 385–388.

Létoffé, A., García-Rodríguez, S.M., Hoppe, S., Canilho, N., Godard, O., Pasc, A., Royaud, I., Ponçot, M. (2019). Switching from brittle to ductile isotactic polypropylene-g-maleic anhydride by crosslinking with capped-end polyether diamine. *Polymer*, 164, 67–78.

Levchenko, I., Ostrikov, K., Zheng, J., Li, X., Keidard, M., Teo, K.B.K. (2016). Scalable graphene production: Perspectives and challenges of plasma applications. *Nanoscale*, 8(20), 10511–10527.

Li, Y., Chen, Q., Xu, K., Kaneko, T., Hatakeyama, R. (2013). Synthesis of graphene nanosheets from petroleum asphalt by pulsed arc discharge in water. *Chemical Engineering Journal*, 215–216, 45–49.

Li, C.Q., Zha, J.W., Long, H.Q., Wang, S.J., Zhang, D.L., Dang, Z.M. (2017). Mechanical and dielectric properties of graphene incorporated polypropylene nanocomposites using polypropylene-graft-maleic anhydride as a compatibilizer. *Composites Science and Technology*, 153, 111–118.

Li, X., Bandyopadhyay, P., Nguyen, T.T., Park, O.K., Lee, J.H. (2018). Fabrication of functionalized graphene oxide/maleic anhydride grafted polypropylene composite film with excellent gas barrier and anticorrosion properties. *Journal of Membrane Science*, 547, 80–92.

Liang, J.Z., Du, Q., Tsui, G.C.P., Tang, C.Y. (2016). Tensile properties of graphene nanoplatelets reinforced polypropylene composites. *Composites Part B: Engineering*, 95, 166–171.

Lotya, M., Hernandez, Y., King, P.J. (2009). Liquid phase production of graphene by exfoliation of graphite in surfactant/water solutions. *Journal of the American Chemical Society*, 131(10), 3611–3620.

Lukes, P., Dolezalova, E., Sisrova, I., Clupek, M. (2014). Aqueous-phase chemistry and bactericidal effects from an air discharge plasma in contact with water: Evidence for the formation of peroxynitrite through a pseudo-second-order post-discharge reaction of H_2O_2 and HNO_2. *Plasma Sources Science and Technology*, 23(1), 015019.

Lux, F. (1993). Models proposed to explain the electrical conductivity of mixtures made of conductive and insulating materials. *Journal of Materials Science*, 28(2), 285–301.

Ma, J., Li, Y., Yin, X., Xu, Y., Yue, J., Bao, J., Zhou, T. (2016). Poly(vinyl alcohol)/graphene oxide nanocomposites prepared by *in situ* polymerization with enhanced mechanical properties and water vapor barrier properties. *RSC Advances*, 6(55), 49448–49458.

Malesevic, A., Vitchev, R., Schouteden, K., Volodin, A., Zhang, L., Van Tendeloo, G., Vanhulsel, A., Van Haesendonck, C. (2008). Synthesis of few-layer graphene via microwave plasma-enhanced chemical vapour deposition. *Nanotechnology*, 19(30), 305604.

McAllister, M.J., Li, J.L., Adamson, D.H. (2007). Single sheet functionalized graphene by oxidation and thermal expansion of graphite. *Chemistry of Materials*, 19(18), 4396–4404.

Miller, S.G., Bauer, J.L., Maryanski, M.J., Heinmann, P.J., Barlow, J.P., Gosau, J.M., Allred, R.E. (2010). Characterization of epoxy functionalized graphite nanoparticles and the physical properties of epoxy matrix nanocomposites. *Composites Science and Technology*, 70(7), 1120–1125.

Notley, S.M. (2012). Highly concentrated aqueous suspensions of graphene through ultrasonic exfoliation with continuous surfactant addition. *Langmuir*, 28(40), 14110–14113.

Novoselov, K.S., Geim, A.K., Mozorov, S.V. (2004). Electric field effect in atomically thin carbon films. *Science*, 306(5696), 666–669.

Novoselov, K.S., Fal'ko, V.I., Colombo, L. (2012). A roadmap for graphene. *Nature*, 490(7419), 192–200.

O'Neill, A., Khan, U., Nirmalraj, P.N. (2011). Graphene dispersion and exfoliation in low boiling point solvents. *Journal of Physical Chemistry C*, 115(13), 5422–5428.

Oxfall, H., Rondin, J., Bouquey, M., Muller, R., Rigdahl, M., Rychwalski, R.W. (2013). Elongational flow mixing for manufacturing of graphite nanoplatelet/polystyrene composites. *Journal of Applied Polymer Science*, 128(5), 2679–2686.

Pan, Y.X., Yu, Z.Z., Ou, Y.C., Hu, G.H. (2000). A new process of fabricating electrically conducting nylon 6/graphite nanocomposites via intercalation polymerization. *Journal of Polymer Science: Part B: Polymer Physics*, 38(12), 1626–1633.

Pang, H., Chen, T., Zhang, G., Zeng, B., Li, Z.M. (2010). An electrically conducting polymer/graphene composite with a very low percolation threshold. *Materials Letters*, 64(20), 2226–2229.

Paredes, J.I., Villar-Rodil, S., Martinez Alonso, A. (2008). Graphene oxide dispersions in organic solvents. *Langmuir*, 24(19), 10560–10564.

Paul, D.R. and Robeson, L.M. (2008). Polymer nanotechnology: Nanocomposites. *Polymer*, 49(15), 3187–3204.

Pol'shchikov, S.V., Nedorezova, P.M., Klyamkina, A.N., Krashenninikov, V.G., Aladyshev, A.M., Shchegolikhin, A.N., Shevchenko, V.G., Sinevich, E.A., Monakhova, T.V., Muradyan, V.E. (2013). Composite materials based on graphene nanoplatelets and polypropylene derived via *in situ* polymerization. *Nanotechnologies in Russia*, 8(1/2), 69–80.

Potts, J.R., Dreyer, D.R., Bielawski, C.W., Ruoff, R. (2011). Graphene-based polymer nanocomposites. *Polymer*, 52(1), 5–25.

Rafiee, M.A., Rafiee, J., Wang, Z., Song, H., Yu, Z.Z., Koratkar, N. (2009). Enhanced mechanical properties of nanocomposites at low graphene content. *ACS Nano*, 3(12), 3884–3890.

Ramanathan, T., Abdala, A.A., Stankovich, S., Dikini, D.A., Herrera-Alonso, M., Piner, R.D., Adamson, D.H., Schniepp, H.C., Chen, X., Ruoff, R.S., Nguyen, S.T., Aksay, I.A., Prud'Homme, R.K., Brinson, L.C. (2008). Functionalized graphene sheets for polymer nanocomposites. *Nature Nanotechnology*, 3, 327–331.

Ray, S.S. and Okamoto, M. (2003). Polymer/layered silicate nanocomposites: A review from preparation to processing. *Progress in Polymer Science*, 28(11), 1539–1641.

Ren, Y., Zhang, Y., Fang, H., Ding, T., Li, J., Bai, S.L. (2018). Simultaneous enhancement on thermal and mechanical properties of polypropylene composites filled with graphite platelets and graphene sheets. *Composites: Part A*, 112, 57–63.

Rosehr, A., Griebe, D., Luinstra, G.A. (2018). Graphite nanosheets–polypropylene composites from in toluene delaminated graphite using atactic polypropylene as dispersant. *Composites Science and Technology*, 156, 28–38.

Rutter, G.M., Crain, J.N., Guisinger, N.P., Li, T., First, P.N., Stroscio, J.A. (2007). Scattering and interference in epitaxial graphene. *Science*, 317(5835), 219–222.

Ryu, S.H. and Shanmugharaj, A.M. (2014). Influence of long-chain alkylamine-modified graphene oxide on the crystallization, mechanical and electrical properties of isotactic polypropylene nanocomposites. *Chemical Engineering Journal*, 244, 552–560.

Salavagione, H.J., Martinez, G., Gomez, M.A. (2009). Synthesis of poly(vinyl alcohol)/reduced graphite oxide nanocomposites with improved thermal and electrical properties. *Journal of Materials Chemistry*, 19(28), 5027–5032.

Samukawa, S., Hori, M., Rauf, S., Tachibana, K., Bruggeman, P., Kroesen, G., Whitehead, J.C., Murphy, A.B., Gutsol, A.F., Starikovskaia, S. (2012). The 2012 plasma roadmap. *Journal of Physics D: Applied Physics*, 45(25), 253001.

Schniepp, H.C., Li, J.L., McAllister, M.J. (2006). Functionalized single graphene sheets derived from splitting graphite oxide. *Journal of Physical Chemistry B*, 110(17), 8535–8539.

Sengupta, R., Bhattacharya, M., Bandyopadhyay, S., Bhowmick, A.K. (2011). A review on the mechanical and electrical properties of graphite and modified graphite reinforced polymer composites. *Progress in Polymer Science*, 36(5), 638–670.

Shanks, R.A. and Cerezo, F.T. (2012). Preparation and properties of poly(propylene-g-maleic anhydride) filled with expanded graphite oxide. *Composites: Part A*, 43(7), 1092–1100.

Shen, J.W., Huang, W.Y., Zuo, S.W., Hou, J. (2005). Polyethylene/grafted polyethylene/graphite nanocomposites: Preparation, structure, and electrical properties. *Journal of Applied Polymer Science*, 97(1), 51–59.

Shioyama, H. (2000). The interactions of two chemical species in the interlayer spacing of graphite. *Synthetic Metals*, 114(1), 1–15.

Shukla, A., Kumar, R., Mazher, J. (2009). Graphene made easy: High quality, large area samples. *Solid State Communications*, 149(17/18), 718–721.

Song, P., Cao, Z., Cai, Y., Zhao, L., Fang, Z., Fu, S. (2011). Fabrication of exfoliated graphene-based polypropylene nanocomposites with enhanced mechanical and thermal properties. *Polymer*, 52(18), 4001–4010.

Stankovich, S., Dikin, D.A., Dommett, G.H.B., Kohlhaas, K.M., Zimney, E.J., Stach, E.A., Piner, R.D., Nguyen, S.B.T., Ruoff, R.S. (2006). Graphene-based composite materials. *Nature*, 442, 282–286.

Stankovich, S., Dikin, D.A., Piner, R.D. (2007). Synthesis of graphene-based nano-sheets via chemical reduction of exfoliated graphite oxide. *Carbon*, 45(7), 1558–1565.

Stauffert, D., Adlerts, J., Aharony, A. (1994). Universality at the three-dimensional percolation threshold. *Journal of Physics A: Mathematical and General*, 27(13), 475–480.

Subrahmanyam, K.S., Panchakarla, L.S., Govindaraj, A., Rao, C.N.R. (2009). Simple method of preparing graphene flakes by an arc-discharge method. *Journal of Physical Chemistry: C*, 113(11), 4257–4259.

Traina, M. and Pegoretti, A. (2012). In situ reduction of graphene oxide dispersed in a polymer matrix. *Journal of Nanoparticle Research*, 14(4), 801–807.

Tsai, H.S., Chih-Chung Lai, C.C., Medina, H., Lin, S.M., Shih, Y.C., Chen, Y.Z., Liang, J.H., Chueh, Y.L. (2014). Scalable graphene synthesised by plasma-assisted selective reaction on silicon carbide for device applications. *Nanoscale*, 6(22), 13861–13869.

Vadukumpully, S., Paul, J., Mahanta, N., Valiyaveettil, S. (2011). Flexible conductive graphene/poly(vinyl chloride) composite thin films with high mechanical strength and thermal stability. *Carbon*, 49, 198–205.

Vallés, C., Drummond, C., Saadaoui, H. (2008). Solutions of negatively charged graphene sheets and ribbons. *Journal of the American Chemical Society*, 130(47), 15802–15804.

Vallés, C., Abdelkader, A.M., Young, R.J., Kinloch, I.A. (2014). Few layer graphene–polypropylene nanocomposites: The role of flake diameter. *Faraday Discussions*, 173, 379–390.

Wallace, P.R. (1947). The band theory of graphite. *Physical Review*, 71(9), 622–634.

Wu, H. and Drzal, L.T. (2013). High thermally conductive graphite nanoplatelet/polye-therimide composite by precoating: Effect of percolation and particle size. *Polymer Composites*, 34(12), 2148–2153.

Wu, C., Dong, G., Guan, L. (2010a). Production of graphene sheets by a simple helium arcdischarge. *Physica E: Low-dimensional Systems and Nanostructures*, 42(5), 1267–1271.

Wu, X., Qi, S., He, J., Duan, G. (2010b). High conductivity and low percolation threshold in polyaniline/graphite nanosheets composites. *Journal of Materials Science*, 45(2), 483–489.

Yang, Y.H., Bolling, L., Priolo, M.A., Grunlan, J.C. (2013). Super gas barrier and selectivity of graphene oxide-polymer multilayer thin films. *Advanced Materials*, 25(4), 503–508.

Young, R.J., Liu, M., Kinloch, I.A., Li, S., Zhao, X., Vallés, C., Papageorgiou, D.G. (2018). The mechanics of reinforcement of polymers by graphene nanoplatelets. *Composites Science and Technology*, 154, 110–116.

Yuan, B., Bao, C., Song, L., Hong, N., Liew, K.M., Hua, Y. (2014). Preparation of functionalized graphene oxide/polypropylene nanocomposite with significantly improved thermal stability and studies on the crystallization behavior and mechanical properties. *Chemical Engineering Journal*, 237, 411–420.

Yue, D.W., Ra, C.H., Liu, X.C., Lee, D.Y., Yoo, W.J. (2014). Edge contacts of graphene formed by using a controlled plasma treatment. *Nanoscale*, 7(2), 825–831.

Zhang, H.B., Zheng, W.G., Yan, Q., Yang, Y., Wang, J.W., Lu, Z.H., Ji, G.Y., Yu, Z.Z. (2010). Electrically conductive polyethylene terephthalate/graphene nanocomposites prepared by melt compounding. *Polymer*, 51(5), 1191–1196.

Zhao, X., Zhang, Q., Chen, D. (2010). Enhanced mechanical properties of graphene-based poly(vinyl alcohol) composites. *Macromolecules*, 43(5), 2357–2363.

Zhao, S., Chen, F., Huang, Y.J., Dong, J.Y., Han, C.C. (2014). Crystallization behaviors in the isotactic polypropylene/graphene composites. *Polymer*, 55(16), 4125–4135.

Zhu, Y., Murali, S., Cai, W., Li, X., Suk, J.W., Potts, J.R., Ruoff, R.S. (2010). Graphene and graphene oxide: Synthesis, properties, and applications. *Advanced Materials*, 22(35), 3906–3924.

2

Morphological Characterization Techniques for Nano-Reinforced Polymers

Adoté Sitou BLIVI[1], Benhui FAN[2], Djimédo KONDO[3] and Fahmi BEDOUI[1]

[1] *Université de technologie de Compiègne, Sorbonne University, France*

[2] *CentraleSupélec, Paris-Saclay University, Gif-sur-Yvette, France*

[3] *Institut Jean Le Rond d'Alembert, Sorbonne University, Paris, France*

Nano-reinforced polymers are considered to be heterogeneous materials for which the heterogeneity is of nanometric size. In this case, their morphological characterization requires the use of experimental tools that are capable of investigating these scales. The microstructure characterization techniques that can meet this requirement are transmission electron microscopy (TEM) and X-ray diffraction (XRD) at small and wide angles. These two observation techniques are complementary. Indeed, TEM ensures a global and direct observation of the microstructure while XRD allows a so-called indirect observation, at smaller scales than TEM. XRD makes it possible to determine characteristic parameters which can be inaccessible to the TEM, such as intermolecular and intramolecular distances.

2.1. Transmission electron microscopy

Among the 2D observation techniques of the microstructure of polymer matrix nanocomposites, TEM seems to be the most suitable. It enables a direct observation

Nanocomposites,
coordinated by Jinbo BAI.

of the microstructure with a sufficiently accurate resolution to determine the size of nanoparticles and aggregates. TEM observations require the preparation of samples that are adapted to nano-reinforced polymers. After observation, image processing is necessary to determine the microstructure parameters.

2.1.1. *Sample preparation and acquisition of the TEM images*

Before any manipulation, the samples used for the TEM observations are taken from the core of the material (the useful part of the specimen). For the observations, the samples are cut into very thin slices with a thickness between 70 and 100 nm in water using a Leica EM UC7 ultramicrotome with a Diatome ultra 45° diamond. The slices are then placed on 300-mesh copper grids and baked at 40°C for 5 min to remove any trace of moisture. The grids are then placed in the microscope for observation. The microscope used is a JEOL-JSM-2100F with an acceleration voltage of 200 kV. During the observation, the slice of material is bombarded by an electron beam that loses energy depending on the density of the area traversed, resulting in a density-dependent gray-scale image on output. The silica nanoparticles have a higher density than the poly(methyl methacrylate) (PMMA); we therefore obtain an image in which the nanoparticles are represented by dark areas (Figures 2.1 and 2.2). Being exposed to the beam can cause the slice to heat, and therefore it is cooled with liquid nitrogen during observation.

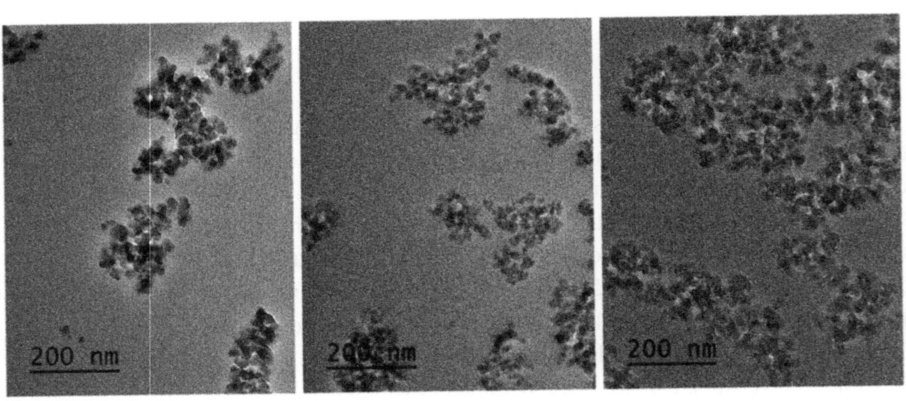

a) b) c)

Figure 2.1. *Snapshots of nanocomposites with 25 nm nanoparticles for volume fractions of a) 2%, b) 4% and c) 6%*

Figures 2.1 and 2.2 represent, respectively, the images of nanocomposites with 25 nm and 150 nm nanoparticles for volume fractions of 2%, 4% and 6%. In these figures, it can be noticed that, for the same scale of observation and the same particle size, the particle density increases with the volume fraction. We can also notice that the dispersion state decreases with the decrease in the nanoparticle size. At a fixed nanoparticle size, the number of nanoparticles increases with increasing volume fraction. By comparing the images in Figure 2.1 with those in Figure 2.2, we can already notice that the dispersion in the case of large particles is better than that of small particles.

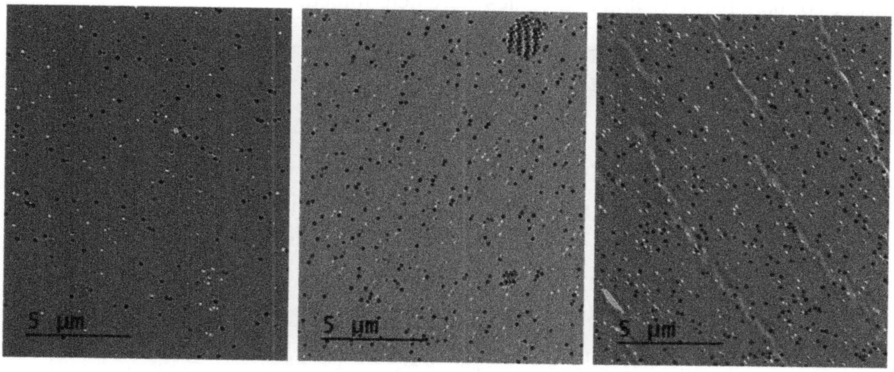

a) b) c)

Figure 2.2. *Snapshots of nanocomposites with 150 nm nanoparticles for volume fractions of a) 2%, b) 4% and c) 6%*

2.1.2. *Size, dispersion and interparticle distance*

The collection of information (nanoparticle and aggregate size, dispersion rate and interparticle distance) on the microstructure of the nanocomposites was done by image processing, using the free software ImageJ (Rasband 2012), of the pictures obtained by TEM. It consists of the removal of the shadow areas by means of uniform illumination of the images, followed by a binarization and an analysis of the particles. This process is achieved in several steps. The goal of the processing is to eliminate the contribution of the matrix and to recover the contribution of the nanoparticles without degrading the contours. In Figure 2.3(a), which corresponds to the original image, one can note that the upper left corner is less illuminated than the lower right corner. The first step of the process consists of removing the shadows and illuminating the image in a uniform way to separate the contrast of the nanoparticles from that of the matrix based on the gray levels (Figure 2.3(b)). This contrast separation is done using the bandpass filter implemented in the ImageJ software. The image is then binarized and the residuals of the matrix are removed by the noise tool (Figure 2.3(c)). At this stage, only the contribution of the nanoparticles remains on the image. The final image (Figure 2.3(d)) is then analyzed for nanoparticle detection. Using the nanoparticle analysis tool available in ImageJ, the surfaces of the detected nanoparticles were recorded, as was the position of their center of mass (coordinates of the center of gravity of the nanoparticle). The detection of the coordinates is done by scanning the image along the x and y axes. On each axis, the scan is done pixelwise. Due to the binarization, all that remains in the image are gray levels that are close to white (255) and those that are close to black (0 nanoparticles). As a result, the contours and surfaces of nanoparticles and aggregates were determined.

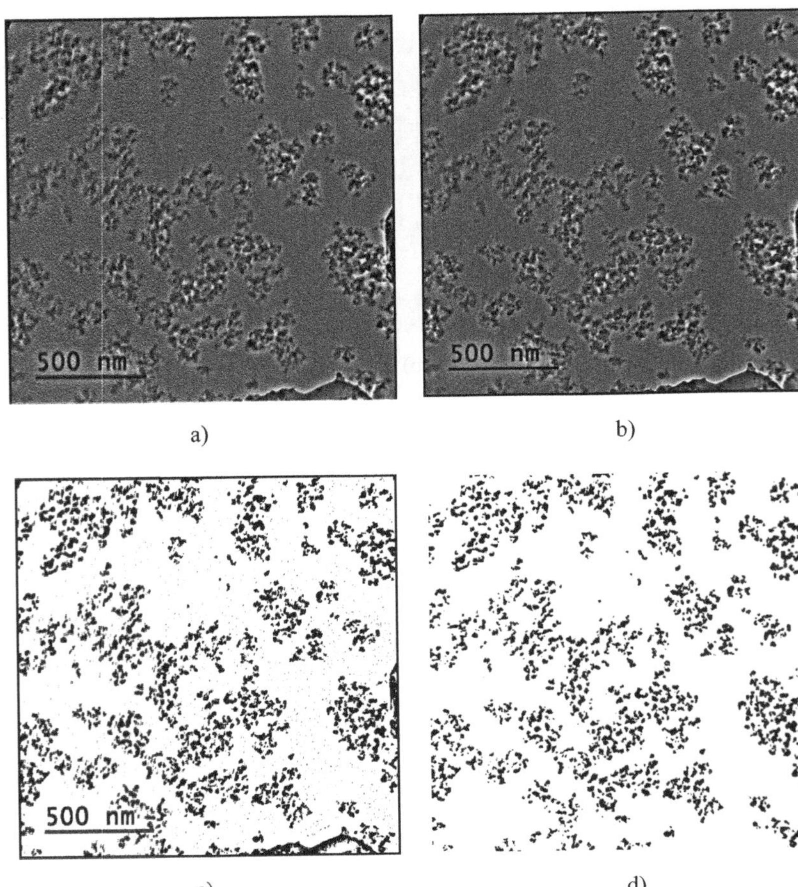

Figure 2.3. *Image analysis with Image J: a) raw image, b) removal of shadow areas and uniform illumination, c) matrix removal and d) removal of the matrix residues*

The surfaces of the nanoparticles were used to calculate their gyration diameter, which we assimilated to the diameter and dispersion rate within the matrix. We wrote a routine in Matlab® to calculate the distance between a nanoparticle and its nearest neighbor by the principle presented in Figure 2.4. The input data of the routine are the positions (coordinates) obtained by the analysis of the particles with the ImageJ software. The principle of the routine consists of calculating the distance between each pair of nanoparticles formed by a nanoparticle of fixed position and another one that can be seen in the picture. It then calculates the difference between the distance of each pair of nanoparticles and the sum of their radii. The interparticle distance is the minimum of these differences. These calculations are repeated for all the nanoparticles present on the plate and for each type of nanocomposite, and the number of plates processed per size is between 6 and 12. For large particles that are well dispersed, six TEM images were sufficient, because they have fairly close dispersion rates. On the other hand, for small particles, up to 12 images were sometimes processed. Increasing the number of images makes it possible to increase the population size and reduce variations from the mean.

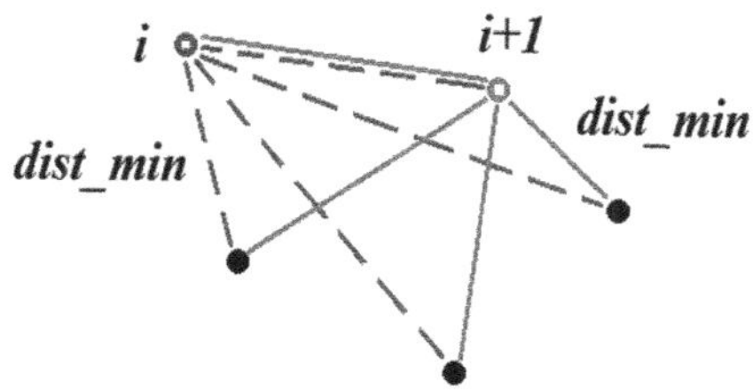

Figure 2.4. *Diagram of the determination of the nearest neighbor distance using the Matlab® code. For a color version of this figure, see www.iste.co.uk/bai/nanocomposites.zip*

The frequencies of nanoparticle size occurrence and interparticle distances are then approximated with a log-normal function. We consider the mean of the log-normal function as the mean of the treated population. The nanoparticle dispersion rate is the percentage corresponding to the mean of the nanoparticle sizes. The size distribution in log-normal function will be used in Chapter 4 for estimating the properties with the micromechanical model.

Figure 2.5 plots the frequency of occurrence (left axis) and the cumulative dispersion rate (right axis) of the 25 nm and 150 nm nanoparticle size for the 2%, 4% and 6% volume fractions. The average nanoparticle size is the center of the log-normal curve (red curve) and the average dispersion rate is the percentage dispersion corresponding to the average. As an example, for the 2%/25 nm nanocomposites, the average size is 26 nm and the corresponding dispersion rate is 67%. "n" is the number of nanoparticles or aggregates used to draw the curve. The analysis of these curves allowed the sizes of the nanoparticles and aggregates to be determined. Therefore, for nanoparticles with theoretical diameters of 25 nm and 150 nm, diameters of 26 nm and 170 nm were measured on average, respectively (Table 2.1). The analysis also helped to determine the size of the aggregates that are particularly present in the case of small nanoparticles. Figure 2.5 also shows the average dispersion rate corresponding to the average size of the nanoparticles. A decrease in the dispersion rate can be observed with the decrease in the size of the nanoparticles. Indeed, for a volume fraction of 2%, the dispersion rate is 67% and 74%, respectively, for 25 nm and 150 nm nanoparticles. This rate is 35% for 15 nm nanoparticles at 2% (see Table 2.1), where aggregates are particularly prevalent. The same observation is made for the other volume fractions. Nanoparticles tend to agglomerate as their size decreases, which reduces the dispersion rate. The agglomeration of small-sized nanoparticles is due to surface tension and their ability to attract each other, which increases as their size decreases (Cheng *et al.* 1993; Banfield and Zhang 2001). The smaller the size of the nanoparticles, the greater the attraction field surrounding them. The choice of combining solution and melt-mixing techniques seems to pay off rather well for dispersing nanoparticles, except for 15 nm nanoparticles. For the other sizes, we obtained more than 60% good dispersion, which is encouraging considering the gap between the nanoparticle scale and the scale at which we work.

By processing the images, it was also possible to determine the interparticle distance, which is the average of the distances between a nanoparticle and its closest neighbor. At constant volume fraction, these distances are a function of the particle sizes. Figure 2.6 shows the distribution of the interparticle distance in nanocomposites with 25 nm and 150 nm diameter nanoparticles for volume fractions of 2%, 4% and 6%. The average interparticle distance is the center of the log-normal curve. The values of the interparticle distances are recorded in Table 2.1. In general, a decrease in interparticle distance is observed with the decreasing size of nanoparticles with fixed volume fraction. A decrease in this distance is also observed when the volume fraction is increased at fixed nanoparticle size. As a matter of fact, at constant volume fraction the number of nanoparticles increases when the size of the nanoparticles decreases. This results in decreasing distances. This decrease in the interparticle distance will reduce the matrix chain mobility and thus lead to a limitation. The limitation will result in changes in the properties of the matrix, which will be discussed in Chapter 3 (Shin *et al.* 2015; Tessema *et al.* 2017).

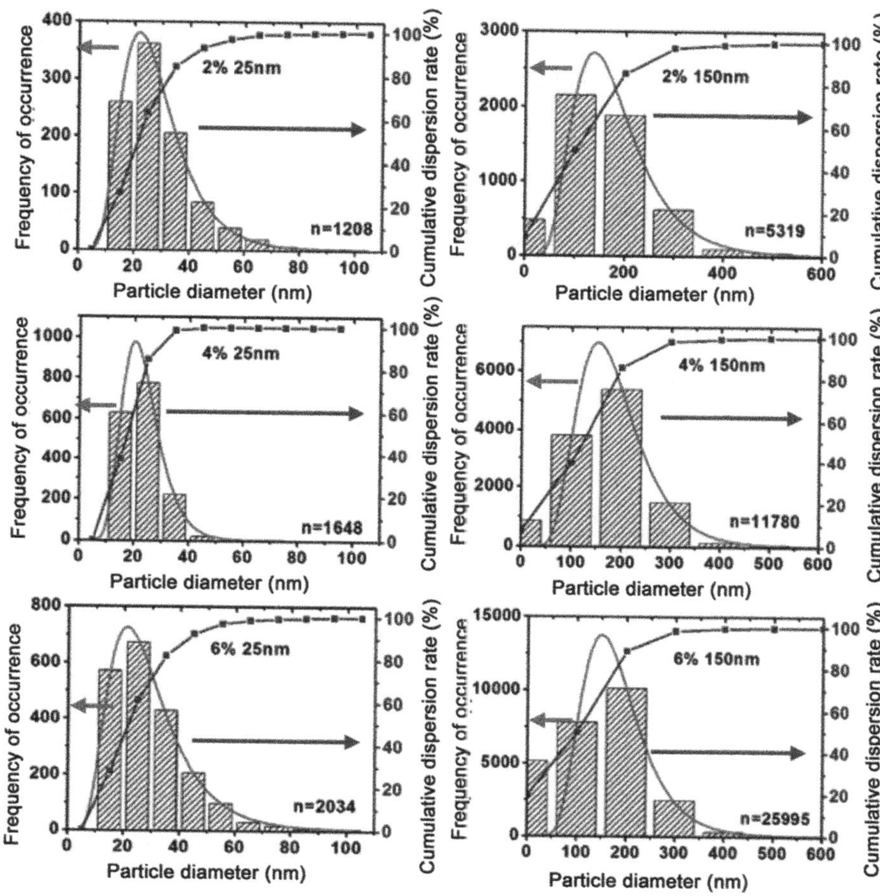

Figure 2.5. *Size distribution and dispersion of 25 nm and 150 nm nanoparticles within the matrix for the volume fractions of 2%, 4% and 6%. For a color version of this figure, see www.iste.co.uk/bai/nanocomposites.zip*

TEM observations have enabled an observation of the microstructure. They first made it possible to verify the spherical size of the fillers and their dimensions. The rate of aggregates present in the microstructure is significant in the case of small size nanoparticles despite the efforts made to disperse them well. The derived information could be used as input in a modeling approach. These observations can also then help to quantify the size effects on the microstructure.

The size effect not only concerns the size of the nanoparticles and aggregates, their dispersion and interparticle distances but also the behavior of the matrix with respect to the size of the nanoparticles. Microstructure parameters related to the nanoparticles were determined through the TEM analyses.

However, the microstructural parameters of the matrix cannot be determined by TEM observations. In order to access them, we will use the experimental technique adapted to the matrix, which is X-ray diffraction.

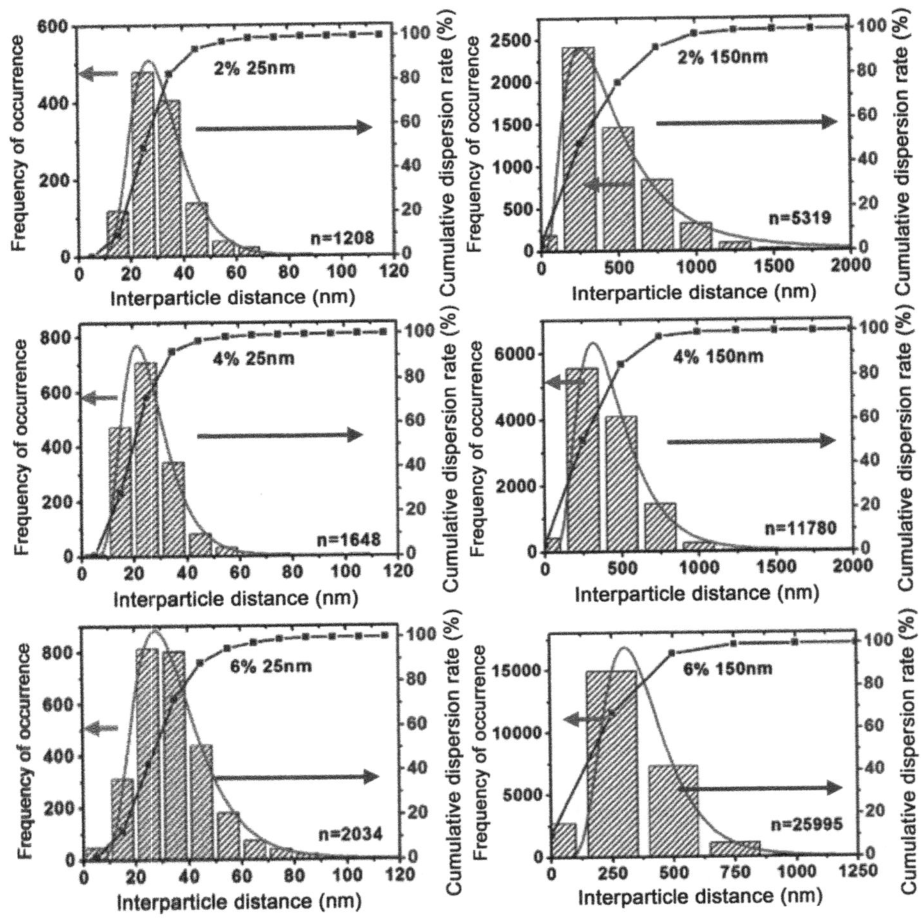

Figure 2.6. *Interparticle distance of 25 nm and 150 nm nanoparticles within the matrix for volume fractions of 2%, 4% and 6%. For a color version of this figure, see www.iste.co.uk/bai/nanocomposites.zip*

Size (nm)	Parameters	2%	4%	6%
15	Diameter (nm)	11.1 ± 0.3	13.8 ± 0.7	11.4 ± 1.2
	Dispersion (%)	35	45	48
	Interparticle distance (nm)	37.9 ± 1.7	27.7 ± 0.2	26.9 ± 0.2
25	Diameter (nm)	26 ± 0.1	22.3 ± 0.1	26.5 ± 0.2
	Dispersion (%)	67	74	65
	Interparticle distance (nm)	30 ± 0.3	24.7 ± 0.16	32.7 ± 0.5
60	Diameter (nm)	65.4 ± 0.5	68.2 ± 0.3	53 ± 0.9
	Dispersion (%)	73	68	50
	Interparticle distance (nm)	47.5 ± 0.1	48.5 ± 0.1	39.7 ± 1.3
150	Diameter (nm)	166.2 ± 1.9	175.6 ± 1.5	171.2 ± 0.8
	Dispersion (%)	74	75	80
	Interparticle distance (nm)	411.7 ± 12.9	403.2 ± 3.6	351.6 ± 0.5
500	Diameter (nm)	452.1 ± 5.4	474.4 ± 9.4	389.6 ± 0.6
	Dispersion (%)	75	63	83
	Interparticle distance (nm)	1,063.8 ± 47.4	825.9 ± 31.2	645 ± 20.7

Table 2.1. *Diameters, dispersion rates and interparticle distances (TEM)*

2.2. X-ray diffraction

The X-ray diffraction measurements performed in this chapter were performed on the synchrotron 5-DND-CAT sector line at the Center for Nanoscale Materials at the Argonne National Laboratory[1]. The wavelength of the beam is $\lambda = 0.7293$ Å^{-1}. In addition to observations at various angles, the equipment available on this line is capable of monitoring temperature for thermal testing.

The measurements at small and wide angles lead to obtaining distances between:

– SAXS (small angle X-ray scattering): $0.00239 \leq q \leq 0.144$ Å^{-1} ($2{,}630 \geq d \geq 44$ Å);

– MAXS (middle angle X-ray scattering): $0.133 \leq q \leq 0.826$ Å^{-1} ($47.2 \geq d \geq 7.61$ Å);

– WAXS (wide angle X-ray scattering): $0.67 \leq q \leq 4.5$ Å^{-1} ($9.4 \geq d \geq 1.4$ Å).

1 Available at: www.anl.gov/cnm.

The advantage of the synchrotron compared to conventional X-ray diffraction lies in the exposure time of the sample to the beam. Due to its high power, the synchrotron is capable of providing a very short exposure time compared to the equipment we have in the laboratory. In addition to the accumulation of noise that can result from a long exposure to the beam, this can also cause a local heating of the material and thus modify the microstructure. The sample exposure and data acquisition times are, respectively, 0.01 s and 0.012 s. The distance between the sample and the SAXS sensor is 8502.5 mm and that to the WAXS sensor is 200 mm.

2.2.1. *SAXS*

Small-angle diffraction measurements are used to determine the size of the nanoparticles, aggregates and interparticle distances. For the tests, the samples were placed in an aluminum crucible that served as a support. The tests were performed on the matrix, the nanoparticles, the nanocomposites and the empty crucible (support). The diffraction data obtained at the end of each test are those for the sample and the crucible. To obtain the data from the sample or the nanoparticles contained in the nanocomposites, we made two corrections. We subtracted the crucible and matrix data first and then applied the Porod correction (Porod 1951; Schnablegger and Singh 2013). The Porod correction is a mean used to correct the curve at large q values of the SAXS data. The curve of the diffracted intensity as a function of the inverse of q^4 (I vs. $1/q^4$) is approximated by linear regression at large values of q. The Porod correction consists of subtracting the value of the intercept of this regression from the diffracted intensity. With the corrected data, we proceeded to analyses in different zones, such as the Guinier zone to determine the size of the nanoparticles, the Porod zone for their shape and the Fourier zone for the interparticle distances.

Figure 2.7 shows the Porod analyses of (a) 60 nm nanoparticle powder and the 4%/60 nm nanocomposite and (b) 150 nm nanoparticle powder and the 4%/150 nm nanocomposite. In Figure 2.7(a), the linear regression of the curves at large values of q showed distributions in q^{-4} and $q^{-3.6}$, respectively, for the nanoparticles in the silica powder and those in the 4%/60-nm nanocomposite. An evolution of the curve in q^{-4} is characteristic of perfectly spherical nanoparticles. It thus reveals that the nanoparticles are spherical before being incorporated into the matrix. An evolution of the curve in $q^{-3.6}$ is characteristic of an intermediate shape between the spherical and flat shapes. Given that the nanoparticles are initially spherical, we deduce that this is the shape of agglomerated nanoparticles. Indeed, in the case of powders, the nanoparticles are isolated from each other and well dispersed, which allows the evolution of the curves at q^{-4}. When the nanoparticles are incorporated into the matrix, chemical bonds are created which, when they are stronger than the nanoparticle–nanoparticle bonds, allow a good nanoparticle distribution within the matrix, but if they are not, they lead to the formation of aggregates. The shape of the aggregates combined with the spherical shape of nanoparticles is responsible for the $q^{-3.6}$ evolution of the 4%/60 nm intensity curve (Schnablegger and Singh 2013; GmbH AP. SAXS-Data Classification 01-Basics).

In Figure 2.7(b), the linear regression at large values of q is q^{-4} for the powder and nanocomposites. We thus notice that, unlike 60 nm nanoparticles, 150 nm nanoparticles are well dispersed in the matrix. In this figure, the periodic evolution of the curve is another observation that gives information on the state of the nanoparticle distribution. This evolution is characteristic of monodispersity (Schnablegger and Singh 2013; GmbH AP. SAXS-Data Classification 01-Basics). The Porod analysis was carried out on nanoparticles and nanocomposites, and we noticed that all powdered nanoparticles present a q^{-4} evolution at large values of q. For nanoparticles smaller than 60 nm in diameter mixed with the matrix, the intensity curve evolves in $q^{-\alpha}$ with $\alpha < 4$. For nanoparticles of diameter 150 nm and above, we have an evolution in q^{-4}, whether they are mixed with the matrix or not. For the same nanoparticle sizes, we also remarked the periodic evolution of the curves, regardless of the volume fraction. We note that the critical size from which the nanoparticles start to agglomerate is between 60 nm and 150 nm.

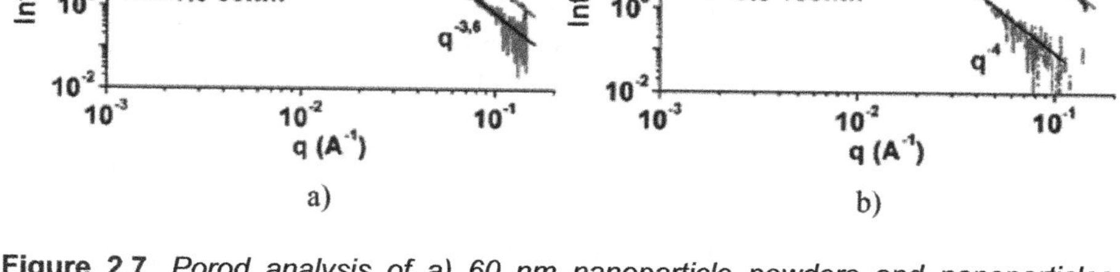

Figure 2.7. *Porod analysis of a) 60 nm nanoparticle powders and nanoparticles contained in the 4%/60 nm nanocomposite; b) 150 nm nanoparticle powders and nanoparticles contained in the 4%/150 nm nanocomposite. For a color version of this figure, see www.iste.co.uk/bai/nanocomposites.zip*

Figure 2.8 shows the Guinier analysis of 60 nm nanoparticles in powder and in the 4% nanocomposite. This analysis makes it possible to determine the average radius of the nanoparticles. It consists of achieving a linear regression at the very low values of the natural logarithm curve of the intensity with respect to the square of the diffusion vector (Ln(I) vs. q^2). Guinier has shown that at very low values of q, the size of nanoparticles can be determined (Guinier and Fournet 1956; Glatter and Kratky 1982). By comparing the linear regression of the nanoparticle powder to that of the nanocomposite, one can note that the regression of the nanocomposite curve is not linear. The nanocomposite has a nonlinear evolution at low values of q, which is characteristic of the presence of aggregates, as observed in Figure 2.7(a). The linear regression is effective when the nanoparticles are well dispersed; otherwise, the analysis can lead to errors, namely by giving the average radius of the set of nanoparticles and aggregates and not the respective radii of the nanoparticles and aggregates. As our materials contain aggregates mainly at low nanoparticle sizes, the Guinier analysis was not conclusive.

We then focused on the deconvolution technique. The first step is to apply the Lorentz correction. This consists of multiplying the diffracted intensity by q^2 and results in bringing out the positions of the peaks, which are sometimes difficult to distinguish. The positions of the peaks, which are determined at maximum intensities, can be used to determine characteristic lengths by the Bragg law. The application of this correction has no effect on the SAXS curves of a randomly oriented crystal or material with well-defined peaks, and the effect is to shift the peak positions slightly to higher values (Dlugosz *et al.* 1976; Cser 2001).

The use of the Lorentz correction is difficult in the case of a multiphase material. Indeed, the diffusion curve of a multiphase material is the sum of the diffusion intensities of the different phases and the interactions between these phases. For example, in a periodic material there are characteristic lengths that, on the SAXS curve, produce a peak or a series of peaks whose exact positions are determined based on the Lorentz correction. If nanoparticles can also be found in the same material, their characteristic scatterings will be combined with those of the characteristic lengths. In this case, with the Lorentz correction, the positions of the peaks with maximum intensities no longer correspond to the characteristic lengths of the periodic phase or to the nanoparticles. A deconvolution of the general curve must therefore be done to separate the different scattering curves and determine the positions of the different peaks. In the case where the Guinier analysis was not conclusive, the deconvolution technique is preferred in order to separate the contributions of the different sizes present in the material.

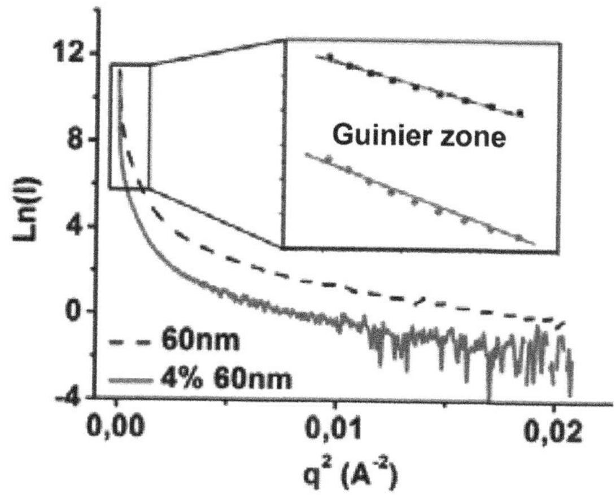

Figure 2.8. *Guinier analysis for powders of 60 nm nanoparticles and nanoparticles contained in the 4%/60 nm nanocomposite. For a color version of this figure, see www.iste.co.uk/bai/nanocomposites.zip*

Figure 2.9 shows the deconvolved curves of the diffracted intensity as a function of the scattering vector of the 4%/15 nm and 4%/60 nm nanocomposites. The solid curves are the experimental curves obtained after the Lorentz correction, and the dashed and mixed curves are those obtained after deconvolution. The dashed curves represent the contributions of the shape factors while the mixed curves are that of the structure factor. The shape factor gives information on the size and areas of the nanoparticles, while the structure factor gives information on the interparticle distances (GmbH AP. SAXS-Data Classification 01-Basics). The peaks of the dashed lines give the sizes of the nanoparticles with the Bragg law and the position of the peak of the dot-dash line curve gives an approximate value of the interparticle distance. The ratio of the area under a form factor curve to the total area is taken as the distribution rate of nanoparticles of corresponding size at this curve. Deconvolution of the 4%/15 nm nanocomposite curve (Figure 2.9(a)) yielded three different sizes of nanoparticles with corresponding ratios. In the case of the 4%/60 nm nanocomposite, two sizes were obtained (Figure 2.9(b)). In the 4%/15 nm nanocomposite, there were 21% nanoparticles with a diameter of 16 ± 1 nm, 60% with 31 ± 3 nm and 19% with 97 ± 2 nm. The 4%/60 nm nanocomposite contained 67% of 63 ± 4 nm diameter nanoparticles and 37% of 123 ± 6 nm diameter nanoparticles. In the two nanocomposites presented, a rate of aggregation can be observed that increases with the decrease in nanoparticle size. By decreasing the size of the nanoparticles, the surface tension increases and the nanoparticles are more attracted to each other and mix less efficiently with the matrix. The position of the peak of the dot-dash line curve gives the average distance of the centers of mass of a nanoparticle and its nearest neighbor. To determine the interparticle distance, the sum of the radii of the two nanoparticles in question must be subtracted from the average distance. In the case of our materials, the variability of the size of nanoparticles and aggregates makes this task more complicated. In order to get an interparticle distance, we subtracted the diameter of the nanoparticles with the highest proportion from the average distance.

Indeed, the higher the proportion of a nanoparticle size, the greater the probability of having a nanoparticle of that size as a nearest neighbor. For example, in the case of the 4%/60 nm nanocomposite, we subtracted 63 nm from the average distance, which is the most represented nanoparticle size in the material (67%). In cases where the nanoparticles are well dispersed, such as for nanoparticles of diameters 150 nm and 500 nm, only one nanoparticle size was obtained and the determination of the interparticle distance is easier. The determination of nanoparticle sizes and interparticle distances was performed for the four volume fractions and five sizes studied, and the results are reported in Table 2.2.

In Table 2.2, we can see that, for small nanoparticle sizes, the size of aggregates increases with the volume fraction and that, unlike what might be expected, their proportion decreases in favor of intermediate-sized nanoparticles (doublet or triplet of nanoparticles). Indeed, by increasing the volume fraction, not only does the probability of having many more aggregates increase but so does that of having many more isolated nanoparticles. Moreover, the nanoparticles agglomerating into a large nanoparticle have a reduced surface tension and, under the effect of the matrix interactions, burst and form aggregates of intermediate size, as can be seen for the 6% and 8% volume fractions.

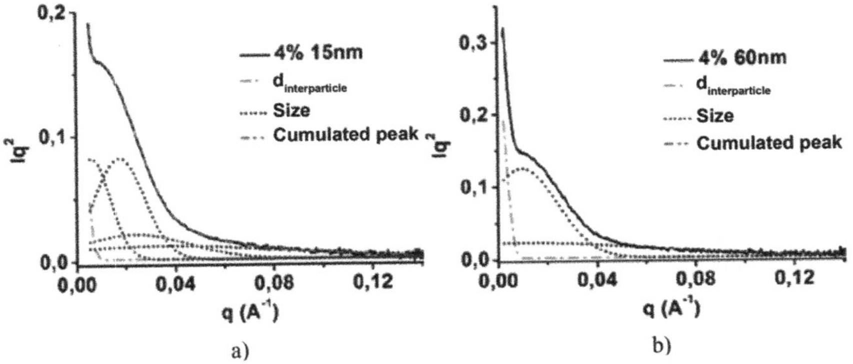

Figure 2.9. *Decomposition of the intensity curves: a) from 4%/15 nm and b) 4%/60 nm after Lorentz correction. For a color version of this figure, see www.iste.co.uk/bai/nanocomposites.zip*

Fraction vol (%)	Theoretical diameter (nm)	Measured diameters (nm)	Proportion (%)	$d_{interparticle}$ (nm) S(q)
2%	15	16 ± 1 (11.1)*	26 (35)*	65
		23 ± 3	15	
		73 ± 2	59	
	25	28 ± 4 (26)*	43 (67)*	61
		86 ± 1	57	
	60	59 ± 2 (65.4)*	30 (73)*	79
		111 ± 4	70	
	150	179 ± 2 (166.2)*	≈100 (74)*	–
	500	492 ± 5 (452.1)*	≈100 (75)*	–
4%	15	16 ± 1 (13.8)*	21 (45)*	86
		31 ± 3	60	
		97 ± 2	19	
	25	26 ± 1 (22.3)*	53 (74)*	152
		84 ± 4	47	
	60	63 ± 4 (68.2)*	67 (68)*	113
		123 ± 6	33	
	150	179 ± 2 (175.6)*	≈100 (75)*	–
	500	492 ± 5 (474.4)*	≈100 (63)*	–
6%	15	16 ± 1 (11.4)*	25 (48)*	–
		27 ± 4	36	
		46 ± 1	22	
		124 ± 7	17	
	25	28 ± 4 (26.5)*	45 (65)*	41
		57 ± 2	55	
	60	61 ± 2 (53)*	64 (50)*	32
		124 ± 5	36	
	150	179 ± 2 (171.2)*	≈100 (80)*	–
	500	492 ± 5 (389.6)*	≈100 (83)*	–

Fraction vol (%)	Theoretical diameter (nm)	Measured diameters (nm)	Proportion (%)	$d_{interparticle}$ (nm)
8%	15	17 ± 1*	31*	114
		30 ± 3	33	
		48 ± 2	29	
		82 ± 5	7	
	25	28 ± 2*	66*	101
		42 ± 4	16	
		95 ± 3	18	
	60	58 ± 1*	31*	105
		109 ± 3	69	
	150	179 ± 2*	≈100*	–
	500	492 ± 5*	≈100*	–

* Diameters and proportions measured by TEM.

Table 2.2. *Diameters, dispersion rates and interparticle distances (SAXS)*

2.2.2. *WAXS*

Wide-angle diffraction tests are performed under the same conditions as those performed at small angles. The only difference is the position of the sensor, which is 200 mm from the sample. While small-angle diffraction helps to determine nanoparticle sizes and interparticle distances, wide-angle diffraction allows intermolecular or intramolecular distances to be determined. The sensor used at wide angles makes it possible to determine distances between 1.4 and 9.4 Å. The values of inter- and intramolecular distances in polymers range between 2 and 7 Å. The range covered by the X-ray wide-angle diffraction will thus enable us to determine these distances and follow their evolution according to the size of the nanoparticles.

Prior to any analysis, the raw data of the matrix, nanoparticles and nanocomposites are corrected and processed. The correction consists of subtracting the signal of the crucible from each signal. Figure 2.10 illustrates the wide-angle diffraction curves of PMMA, the 60 nm diameter silica nanoparticles and the 4%/60 nm nanocomposite. The silica curve shows a broad peak around $q = 1.555$ $Å^{-1}$. The broad peaks are characteristic of amorphous materials and the silica diffraction curve confirms the choice of the silica structure. On the diffraction curve of the matrix three broad peaks can be observed, which also attest to the amorphous structure of the matrix. The same peaks are also observed for the 4%/60 nm nanocomposite. In addition to the matrix peaks, one can also note, although not visible, the silica nanoparticle peak on the 4%/60 nm nanocomposite curve. Furthermore, compared to the PMMA curve, we observe a slight variation in the slope of the 4%/60 nm curve around the position of the silica peak. This variation is due to the silica signal that is included in the global signal of the nanocomposite. This signal, when well determined, is used for the calculation of the mass fraction of nanoparticles present in the material. The mass fraction is obtained by the ratio of the area under the curve of the nanoparticles to the total area of the nanocomposite.

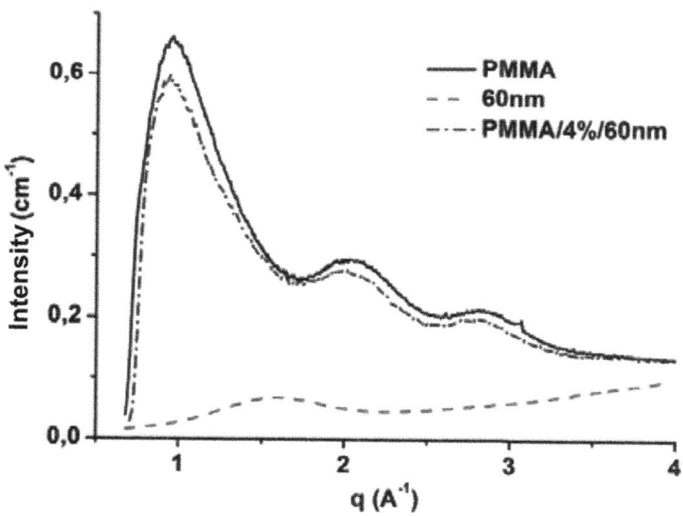

Figure 2.10. *Diffraction PMMA, 60 nm nanoparticles and 4%/60 nm nanocomposite curves. For a color version of this figure, see www.iste.co.uk/bai/nanocomposites.zip*

On the one hand, these observations confirm that the addition of nanosized fillers did not change the structure of the matrix (amorphous) and, on the other hand, validate the choice of our materials. It can thus be deduced that any variation of the characteristic lengths of the matrix in the presence of nanoparticles will only be due to the size of the latter.

To get the characteristic lengths in the matrix and nanocomposites, we applied a method that consists of deconvoluting the corrected experimental signal into Gaussians (Wang *et al.* 2008; Gelineau 2015). Each Gaussian corresponds to an intramolecular or intermolecular distance. The deconvolution was done with a baseline connecting the two ends of the signal (Wang *et al.* 2008; Gelineau 2015). Figure 2.11 represents the deconvoluted signal of the matrix. The matrix curve shows three peaks. The deconvolution of this curve could be done with three Gaussians, however, the asymmetry of the first peak may be the result of the superposition of two peaks. Further works in the literature have shown that the wide-angle diffraction curve of PMMA consists of four peaks (Lovell and Windle 1981). The deconvolution of the curve is done with four Gaussians. Peak 1 corresponds to the intermolecular distance and the other three correspond to intramolecular distances. As a matter of fact, the position of peak 1 (q_1 = 0.916 Å^{-1}) is used to determine the distance between the chains of the matrix and those of peaks 2 (q_2 = 1.125 Å^{-1}), 3 (q_3 = 2.05 Å^{-1}) and 4 (q_4 = 2.854 Å^{-1}) to determine the distances between the different groups which constitute the matrix chains. In their work, Lovell and Windle (1981) attributed the characteristic length corresponding to the position of peak 2 to the distance between two consecutive planes of carbonyl groups. Since the PMMA utilized in this research is atactic, similar to the one whose conformation is proposed by Lovell and Windle (1981), the plane defined by the carboxyl groups is the one which passes through two groups located on each side of the carbon chain. The distance d_3 corresponds to the distance between two planes of the methyl groups and d_4 is the distance between backbone carbon atoms (Lovell and Windle 1981). Using Bragg's law, we obtain d_1 = 6.86 Å, d_2 = 5.59 Å, d_3 = 3.06 Å and d_4 = 2.2 Å. These distances are similar to those found in the literature.

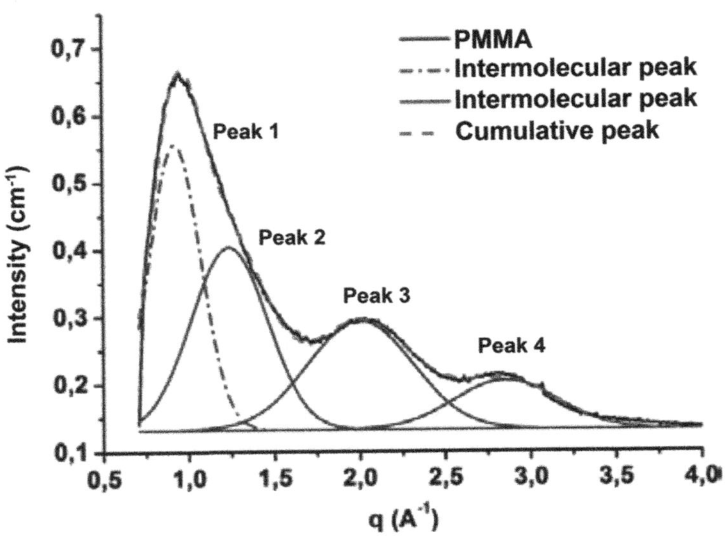

Figure 2.11. *Deconvolution of the of PMMA wide-angle diffraction. For a color version of this figure, see www.iste.co.uk/bai/nanocomposites.zip*

After determining the characteristic lengths existing in the matrix, we performed a comparison of the matrix diffraction curve with those of these nanocomposites as a function of size. In order to compare curves that are at the same scale, we normalized each diffraction curve by its maximum intensity. Figure 2.12 shows the PMMA diffraction curves and its 4% nanocomposites for the different sizes at a temperature of 30°C. Only the areas concerning the first three peaks of the matrix are shown because no significant variation was observed for the position of peak 4. A difference in the shape of the curves is observed in the area corresponding to the positions of peak 2 and the silica peak that we will name peak 5. The position of peak 3 does not seem to change. This difference in shape is reflected by a swelling of the curve. This swelling becomes more and more important with the decrease in the size of the nanoparticles. It may be due to an excessive proportion of nanoparticles in the matrix or to a shift in the positions of peaks 2 and/or 3 toward large values of q. Since the volume fraction of the nanoparticles is constant, it is deduced that this swelling is due to a shift in the positions of peaks 2 and/or 3 toward large values of q. This shift is all the more important as the size of the nanoparticles decreases. The more the position of the peaks is shifted toward higher values of q, the larger the corresponding values by the Bragg law. We also note that the position of the peak with maximum intensity, that corresponds to that of peak 1, varies very little with the size of the nanoparticles. This study shows that at the temperature of 30°C (in the glassy state), the size effect of the nanoparticles on the characteristic lengths of the matrix results in a shift in the position of the peaks corresponding to the distances d_1 and/or d_2 and, more particularly, of d_2. The size effect leads to a decrease in the interparticle distance and the intramolecular distance between the carbonyl planes.

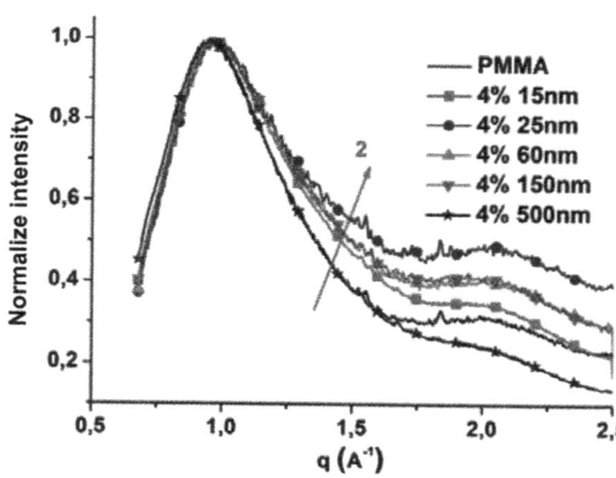

Figure 2.12. *Comparison of the wide-angle diffraction curves of PMMA and its nanocomposites loaded to 4% and different sizes. For a color version of this figure, see www.iste.co.uk/bai/nanocomposites.zip*

In order to achieve a quantitative study of the size effect on the micro-structural parameters of the matrix, the diffraction curves of the nanocomposites were deconvolved and the corresponding distances to the peak positions are compared. Considering the high sensitivity of d_2 to the variation of the nanoparticle size compared to the sensitivity of d_1, we compared the distance d_2 of the matrix to that of the nanocomposites. Figure 2.13 presents the evolutions of the d_2 distances of the nanocomposites as a function of the nanoparticle size for the four volume fractions. The d_2 distances of the nanocomposites are normalized by that of the matrix (d_2 = 5.59 Å). It is generally observed that the intramolecular distance separating the carbonyl planes in the nanocomposites is smaller than that present in the matrix except for 4%/500 nm, 2%/150 nm and 2%/500 nm. At constant volume fraction, d_2 decreases with decreasing nanoparticle size independently of the volume fraction. The nanoparticles with their nanometric dimensions mix with the matrix chains. The nanometric dimensions of the nanoparticles seem to have an effect on the conformation of the matrix chains. In response to the presence of nanoparticles in the matrix, the chains reorganize themselves by changing their configuration.

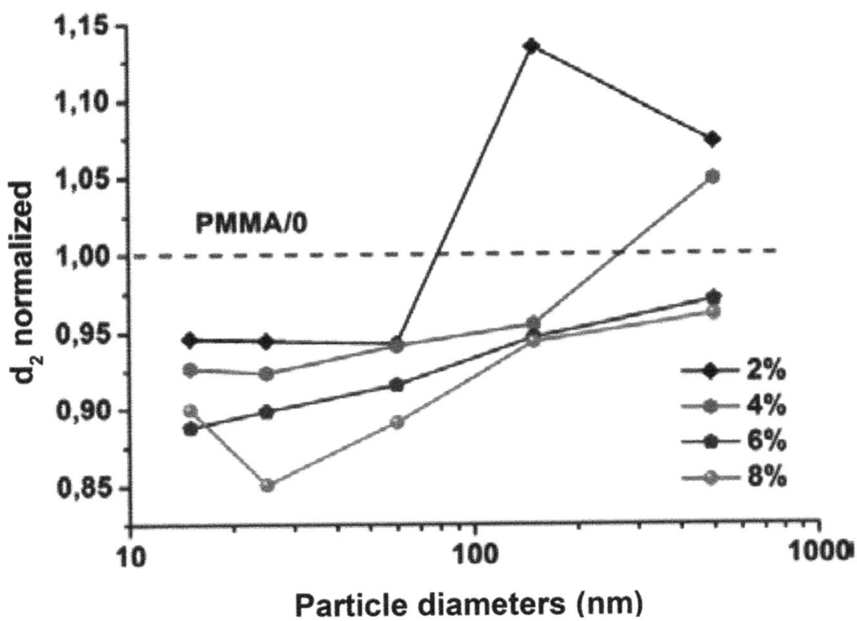

Figure 2.13. *Evolution of the normalized intramolecular distance (d_2) of the matrix in nanocomposites as a function of nanoparticle size and volume fraction (d_2 PMMA = 5.59 Å). For a color version of this figure, see www.iste.co.uk/bai/nanocomposites.zip*

X-ray diffraction was used to determine the characteristic lengths of the nanoparticles (SAXS) and the matrix (WAXS). With the SAXS observations, the size of the nanoparticles and aggregates, as well as their rate, were determined with much more precision. The TEM also led to determining these parameters but it was less easy to grasp the size of the aggregates and their rate at low nanoparticle sizes. The WAXS observations made characteristic lengths that are not visible with TEM accessible. Thus, they showed that the inter- and intramolecular bonds are dependent on the size of the nanoparticles. The decrease in nanoparticle size is accompanied by a decrease in the length of these bonds.

2.3. Conclusion

The study of the effect of filler size on the properties of nanocomposites requires a good choice of matrix, fillers, processing and also a fine microstructure characterization. The main criterion, which is to have the size of the fillers as the only variable factor among the parameters that can modify the microstructure, led to the choices of PMMA for the matrix and silica nanoparticles for the fillers. These choices have allowed us to overcome any crystallization of the matrix in the presence of the fillers, as shown by the diffraction curves of nanocomposites.

The different observation techniques (TEM and SAXS) revealed a variation of the characteristic lengths according to the size of the nanoparticles and at all scales. At the microscopic scale, direct observations showed that the choice of the implementation process allowed for a good distribution of the nanoparticles in the matrix but was not sufficient to avoid the aggregation of small nanoparticle sizes. This aggregation, even if it does happen, is limited to the formation of nanoparticle doublets or triplets, which is still acceptable given the strong attractiveness that exists between silica nanoparticles. The interparticle distances also decrease with decreasing nanoparticle size. At the molecular scale (WAXS), the size of the nanoparticles acts on the interchain bonds by confining the chains between the nanoparticles. This results in a decrease of these bonds with the decrease in the nanoparticle sizes. The decrease in these distances will lead to a reduction of chain mobility. As the properties of the polymers are intrinsically linked to chain mobility, this reduction in the distance could generate a modification of the properties of the matrix.

Morphological characterization has made it possible to identify parameters from the microstructure up to the molecular scale (particle and aggregate sizes, interparticle and interchain distances). In Chapter 3, we will focus on the impact of size on the physicochemical and mechanical properties and we will try to link the identified parameters to the measured properties.

2.4. References

Banfield, J.F. and Zhang, H. (2001). Nanoparticles in the environment. *Rev. Mineral Geochemistry*, 44(1), 1–58.

Cheng, B., Kng, J., Luo, J., Dong, Y. (1993). Relation of structure stability and Debye temperature to crystal size of nanometer sized TiO_2 powders. *Chin. J. Mater. Res.*, 7(3), 240–243.

Cser, F. (2001). About the Lorentz correction used in the interpretation of small angle X-ray scattering data of semicrystalline polymers. *J. Appl. Polym. Sci.*, 80, 2300–2308.

Dlugosz, J., Fraser, G.V., Grubb, D., Keller, A., Odell, J.A., Goggin, P.L. (1976). Study of crystallization and isothermal thickening in polyethylene using SAXD, low frequency Raman spectroscopy and electron microscopy. *Polymer*, 17(6), 471–480.

Gelineau, P. (2015). Caractérisation morphologique et homogénéisation élastique et visco-élastique de polymère. PhD Thesis, Université de technologie de Compiègne.

Glatter, O. and Kratky, O. (1982). *Small Angle X-ray Scattering*. Academic Press, London.

González, I., Eguiazábal, J.I., Nazábal, J. (2008). Effects of the processing sequence and critical interparticle distance in PA6-clay/mSEBS nanocomposites. *Eur. Polym. J.*, 44(2), 287–299.

Guinier, A. and Fournet, G. (1956). Small angle scattering of X-rays. *J. Polym. Sci.*, 19, 594–594.

Kavesh, S. and Schultz, J.M. (1970). Lamellar and interlamellar structure in melt-crystallized polyethylene I. Degree of crystallinity, atomic positions, particle size, and lattice disorder of the first and second kinds. *J. Polym. Sci., Part A-2: Polym. Phys.*, 8(2), 243–276.

Lovell, R. and Windle, A.H. (1981). Determination of the local conformation of PMMA from wide-angle X-ray scattering. *Polymer*, 22(2), 175–184.

Porod, G. (1951). Die Röntgenkleinwinkelstreuung von dichtgepackten kolloiden Systemen. *Kolloid-Z*, 124, 83–114.

Rasband, W. (2012). ImageJ software release 1.48v [Online]. Available at: http://imagej.nih.gov/ij.

Schnablegger, H. and Singh, Y. (2013). The SAXS guide: Getting acquainted with the principles. Guide, Anton Paar GmbH, Austria, 37–84.

Shin, H., Yang, S., Choi, J., Chang, S., Cho, M. (2015). Effect of interphase percolation on mechanical behavior of nanoparticle-reinforced polymer nanocomposite with filler agglomeration: A multiscale approach. *Chem. Phys. Lett.*, 635, 80–85.

Tessema, A., Zhao, D., Moll, J., Xu, S., Yang, R., Li, C., Kumar, S.K., Kidane, A. (2017). Effect of filler loading, geometry, dispersion and temperature on thermal conductivity of polymer nanocomposites. *Polym. Test.*, 57, 101–106.

Wang, W., Murthy, N.S., Chae, H.G., Kumar, S. (2008). Structural changes during deformation in carbon nanotube-reinforced polyacrylonitrile fibers. *Polymer*, 49, 2133–2145.

3

Size Effects on Physical and Mechanical Properties of Nano-Reinforced Polymers

Fahmi BEDOUI[1], Adoté Sitou BLIVI[1], Benhui FAN[2] and Djimédo KONDO[3]

[1] Université de technologie de Compiègne, Sorbonne University, France

[2] CentraleSupélec, Paris-Saclay University, Gif-sur-Yvette, France

[3] Institut Jean Le Rond d'Alembert, Sorbonne University, Paris, France

The interaction of particles with the matrix will be quantified at different scales. It should be noted that since the volume fraction of the different samples was kept constant, the effects observed will then be attributed to the effect of size. This effect will be investigated on different aspects – mechanical, physical and thermal – in order to understand the effect of the particle size across the existing scales in the material.

3.1. Size effect on the glass transition temperature

3.1.1. *Differential scanning calorimetry*

Differential scanning calorimetry (DSC) is a technique used to study the thermal properties of materials, including glass transition and decomposition temperature.

For a color version of all the figures in this chapter, see www.iste.co.uk/bai/nanocomposites.zip.

Nanocomposites,
coordinated by Jinbo BAI. © ISTE Ltd 2022.

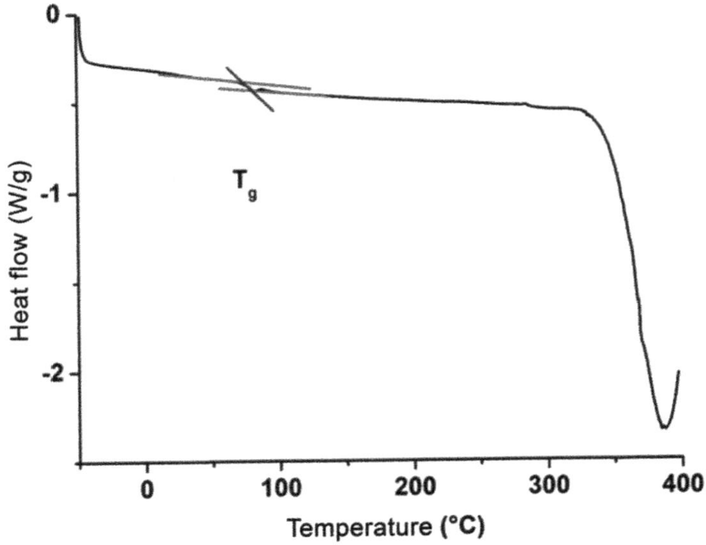

Figure 3.1. *DSC curve of 4%/60 nm nanocomposites*

The tests were performed on a DSC Q100 machine with hermetic aluminum crucibles under a nitrogen atmosphere to avoid any reaction with the external environment. The samples, which were previously baked at 40°C, were subjected to two heating cycles. The first cycle, which goes from –50°C to 200°C with a heating rate of 10°C/min, and then to –50°C, makes it possible to eliminate the residual traces of moisture and solvent that could pollute the measurements. The measurement of the temperature of the vitreous transition (T_g) is made during the second cycle carried out between –50°C and 400°C (limit temperature of the machine) with a speed of 10°C/min. Figure 3.1 shows the DSC curve of the 4%/60 nm nanocomposite. Three transformation zones can be observed on the curve. These transformations occur in the matrix because the temperature of the first transformation of silica is about 600°C. The three observed transformations correspond to the secondary (≈ 25°C), vitreous (≈ 80°C) and decomposition (≈ 387°C) transition zones of the matrix, respectively. The secondary transition is due to local movement of the matrix chains. The vitreous transition is that of the cooperative movement of the chains. The absence of a crystallization peak is clearly visible and shows that the matrix retains its amorphous structure even with nanoparticles of nanometric size. The effect of size on the glass transition temperature will be the point of focus of this work. The determination of the T_g value is carried out using the TA Universal Analysis software provided by the manufacturer of the DSC machine. The procedure consists of indicating the start

(Onset) and end (Offset) points of the transition zone and recovering at the output the corresponding T_g value. The tests were performed three to six times for each type of material and the result presented in this chapter is the average thereof.

Figure 3.2 shows the glass transition zones (DSC) of the 4% filled nanocomposites. It can be observed that between the 500 nm and 15 nm curves, there is a slight shift of the transition zone toward higher temperature values. The T_g values are determined and averaged over three trials.

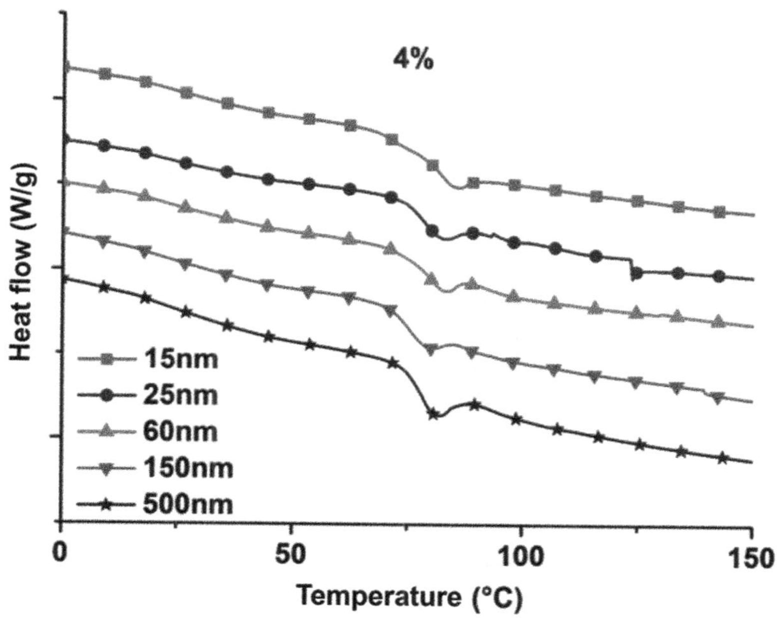

Figure 3.2. *DSC curves: glass transitions zones of 4% filled nanocomposites*

3.1.2. *Dynamic mechanical analysis*

The tests were performed on a DMA$^+$ 150 machine from Metravib with a 150 N force cell. It can perform tensile, compression, shear tests, etc., by controlling the strain rate and the temperature of the specimen. The tests are done in traction-compression without imposing static strain but only a dynamic deformation of 5×10^{-4}. The frequency is 1 Hz and the studied temperature range is from 20°C to 120°C, with a heating rate of 2°C/min.

Figure 3.3 represents the tanδ curves of the 4% filled nanocomposites of the different nanoparticle sizes. The vitreous transition temperature corresponds to the temperature of the maximum tanδ. Two observations can be made with the decrease in the nanoparticle size. There is a decrease in the maximum tanδ value and an increase of the T_g. The decrease in the maximum of tanδ is reflected in the material by a decrease in its dissipative properties with respect to those of conservation $\tan\delta = E''/E'$. It follows that the decrease in the size of the nanoparticles hardens the material. The same effect is visible on the thermal properties, where an increase in the T_g is observed with the decrease in the size of the nanoparticles. For each type of material, the T_g values are determined and averaged.

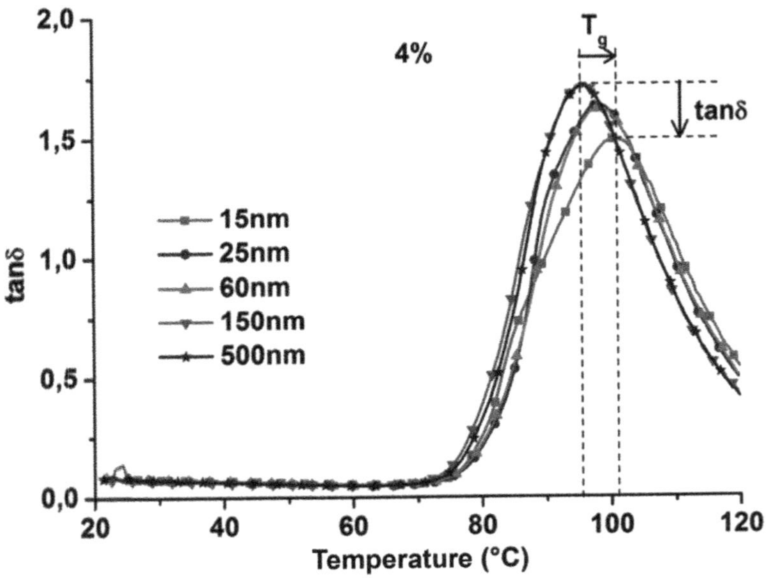

Figure 3.3. *Damping factor (tanδ) curves of 4% filled nanocomposites*

3.1.3. *Wide-angle temperature synchrotron X-ray diffraction (WAXS)*

X-ray diffraction (XRD) is a physicochemical analysis technique that consists of sending a beam onto a sample and recovering, through detectors, the different wavelengths (λ) that have reacted with the elements of the sample. The value of the wavelengths is used to determine the position of the elements that have interacted with the beam, and also to calculate the distance that separates them by the following relations:

$$\nu^{j} = 0.17 \quad [3.1]$$

$$d = \frac{\lambda}{2 \cdot \sin(\theta)} = \frac{2 \cdot \pi}{q} \quad [3.2]$$

This test, when performed in the laboratory, requires a significant exposure time to the beam, which can cause local heating in polymers that have a very temperature sensitive microstructure. This heating can locally modify the microstructure, especially at high temperatures where chain mobility is more important. Instead of conventional X-ray diffraction, we have opted for synchrotron X-ray diffraction tests, which feature a power that allows for very short exposure times (less than a second) and for making observations while varying the temperature of the sample (Ryan *et al.* 1997; Barnes *et al.* 2000). The tests were performed on the 5-ID-DND line of the synchrotron at the Argonne[1] National Laboratory.

During the test, the sample is placed in an aluminum crucible, which is held in a DSC cell through which the beam passes. The DSC tests are performed between 30 and 150°C with a heating rate of 10°C/min. We performed observations with a step of 1°C and a beam exposure time of 0.005 s. The tests are performed with the empty crucible, the different nanoparticle sizes and all nanocomposites. For each material, the data are grouped (Figure 3.4) and processed to compare the evolution of the diffraction curves (intensity versus diffusion vector) as a function of temperature. The treatment consisted of subtracting from the sample data those of the crucible and the DSC cell used during the test. The crucible and the cell also diffract and it is important to subtract their data to have the diffraction data of the content (nanoparticles or nanocomposites).

Figure 3.4 shows the evolution of the corrected wide-angle diffraction curves of the 4%/25 nm nanocomposite as a function of temperature. In the figure, we can note, independently of the temperature, the peaks presenting a diffuse halo characteristic of an amorphous microstructure. It shows that the matrix remained amorphous in the presence of nanoparticles and at high temperatures. An increase in the value of the maximum intensity can be observed with the increase in temperature. A shift is also observed in the position of the peak at maximum intensity toward low values of q when the temperature increases. At 30°C, the position of the peak at maximum intensity is around q = 0.96 $Å^{-1}$ while it is at q = 0.86 $Å^{-1}$ when the temperature is 150°C. All of these observations are indicators of change occurring in the material and can be used to determine the vitreous transition temperature (Nguyen 2014; Gelineau 2015). The silica used has its first transition point above 600°C, which is to say that all of these observations are relative to the microstructure of the matrix only. We shall use this information to

1 See: www.anl.gov/com.

determine the temperature zones where the matrix changes behavior. The process utilized (Ryan *et al.* 1997; Gelineau 2015) consists of decomposing the experimental signal into Gaussians, in order to distinguish the contribution of the matrix from that of the nanofillers. The transition zones are then targeted by summing the area of each of the matrix Gaussians and their variation with respect to temperature. However, in the case of most of our materials, this technique did not allow us to obtain a variation of the area under the matrix curves with respect to temperature, thus making it difficult to determine the transition zone. The same observation was also made in the work of Gélineau (2015). This does not mean that the transition zone is non-existent. The baseline used at the time of deconvolution could have an effect on the area of the Gaussian. As a matter of fact, the starting and ending points of the signal change with temperature, making the baseline dynamic. We believe that this change in baseline with temperature affects the area under the curves. In order to avoid this problem, we are interested in the evolution of the characteristic lengths as a function of temperature through the position of the peaks that does not depend on the baseline.

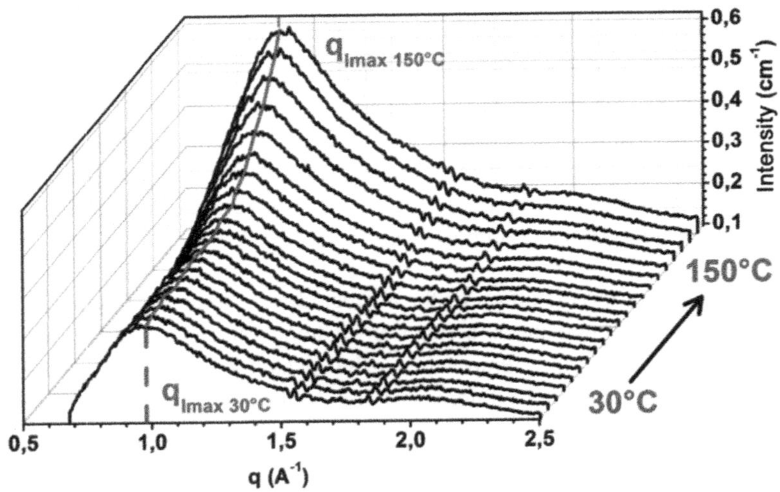

Figure 3.4. *Diffraction patterns versus temperature of the 4%/25 nm nanocomposite*

Polymeric materials are very sensitive to temperature. This sensitivity is reflected in microstructural changes, and more particularly in the characteristic lengths. Nevertheless, the chain mobility in the polymer does not occur at the same temperature. Indeed, polymers have different types of transitions, such as tertiary or gamma (γ), secondary or beta (β) and principal or alpha (α), which occur at different temperatures and mobilize different types of chains. The volume of polymers changes under the action of temperature. This change in volume is due to an evolution in intermolecular distances. The intermolecular distances represent the lengths of the weak interactions (Van der Waals) that are the first to be stressed in the polymer. To determine the main transition temperature of our materials by the temperature DRX test, we followed the evolution of the interchain distance according to temperature. The diffraction curves obtained according to temperature were deconvoluted and the position of peak 1 was located. The corresponding distances are calculated using Bragg's law. Figure 3.5 represents the evolution of the intermolecular distance (d_1) according to temperature for nanocomposites filled with 4% of particles of sizes 15 nm, 60 nm and 500 nm. The temperature values are on the abscissa and the intermolecular distance on the ordinate.

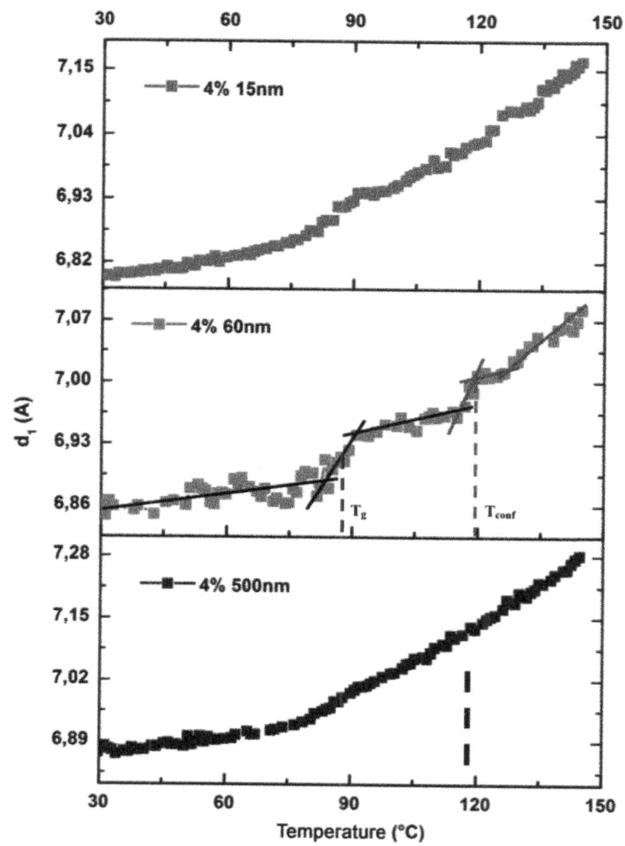

Figure 3.5. *Intermolecular distance versus temperature for the 4% nanofilled matrix of 15 nm, 60 nm and 500 nm particle sizes*

Then, for each size, an increase of d_l can be noted with the increase in the temperature. This evolution can be classified in different zones: stability zones where the variation of d_l is very progressive and others of instability where the slope changes. The increase in the slope results from a greater chain mobility at the corresponding temperature values. The vitreous transition is an area where all of the chains of the matrix acquire a greater mobility. We consider the areas of rapid change in the slope as transition zones and the average of the starting and ending temperatures of the zone as the transition temperature. On these curves, two transition zones were observed. The first one is close to the glass transition zone observed in DSC and the second is at a higher temperature. For the first transition zone, the transition temperature determined is not a vitreous transition temperature in the usual sense. However, given its proximity to that measured in DSC, we shall assimilate it to the T_g. The second transition zone which was not visible in DSC can be the transition zone of the confined chains. These chains have a reduced mobility compared to the other chains because of their confinement in the subsurface of the particles. We note this temperature T_{conf} to designate the confinement. As in the DSC and dynamic mechanical analysis (DMA) cases, the values of the T_g are indicated in Table 3.1 and represented in Figure 3.6. The T_{conf} values are also recorded in Table 3.1 and their evolution with particle size is shown in Figure 3.7.

In Figure 3.6, the curves of the T_g of the PMMA/0 matrix and of the nanocomposites are grouped according to the size of the nanoparticles for the four 2% and 4% volume fractions. These curves are obtained by using the three different thermal analysis techniques (DSC, DMA and WAXS-DSC). The mixed two-dot and long dashed lines at the bottom of each curve represent the T_g values of PMMA/0 obtained in DSC and DMA, respectively. It can be seen that the trend of the curves is the same from one volume fraction to another, even though the analysis technique is different. An increase in the T_g value is thus observed with the decrease in the nanoparticle sizes (size effect). This trend can be broken down into two stages: a first stage between 500 nm and 60 nm where the increase is small, and a second between 60 nm and 15 nm where the increase is more pronounced. In Figure 3.6, we notice that the T_g of the nanocomposites is higher than that of the matrix, regardless of the analysis technique. This may be due to different phenomena such as the hydrogen bonds that are created between the matrix and the nanoparticles (Wang *et al.* 1991; Tripp and Hair 1992). Due to the -OH groups existing on their surface, the silica nanoparticles create hydrogen bridges with the carbonyl groups of the matrix, thus making the nanocomposite matrix less sensitive to temperature compared to the matrix alone.

The two-phase increase in T_g between nanocomposites, as explained above, is due to the hydrogen-nanoparticle-matrix bonds and the size of the nanoparticles.

Indeed, the number of -OH groups present on the surface of nanoparticles increases with the decrease in the nanoparticle size (Tripp and Hair 1992). The interest in using spherical nanoparticles comes from their large area-volume ratio, and this ratio is even greater as the size of the nanoparticles decreases. Therefore, large nanoparticles, although better dispersed than small nanoparticles, present a lower surface-to-volume ratio than small nanoparticles and form fewer hydrogen bonds with the matrix chains. The size effect thus becomes more visible at the size where the interactions are preponderant in the nano-composite, and this critical size seems to be between 60 nm and 100 nm in our study. This critical size region is the same as the one observed in the morphological characterization. The interparticle distance also decreases with the decrease of the particle size, which leads to a confinement of the chains. The confined area increases with the decrease of the size. Another advantage of small nanoparticles resides in their small size. This small size allows them to better adhere to the matrix. They form a homogeneous system with the matrix and reinforce their interactions, which increases the T_g value. Obtaining such a mechanism requires a good nanoparticle dispersion (2% and 4% cases). On the other hand, their small size increases their reactivity and when they are too close to each other (poorly dispersed), they tend to attract themselves and agglomerate. In this case, they reduce their bonding surface with the matrix and the T_g value drops or varies little. This is the case for the 6% volume fraction where it was observed, with the three techniques used, that between 60 nm and 15 nm the curve falls or forms a plateau. In the case of the theoretical 8% fraction, the difference between the volume fractions measured for the different particle sizes – 11.7 ± 2.19% for 15 nm particles and 8.59 ± 0.2 for 150 nm particles (Table 3.1), namely a difference of 3.11% – does not allow the increase in T_g to be attributed to the size of the particles. This increase is also due to a reinforcing effect because of the 3.11% difference in volume fractions.

The nanoparticles, due to the strong interactions that they establish with the chains that surround them, prevent the transition from occurring at the same temperature as the material in the *bulk*. Indeed, contrary to the chains located in the matrix, the confined ones need more energy (calorific) to break the hydrogenic bridges that they establish with the nanoparticles before starting their mobility. It takes place at staggered temperatures, which explains the appearance of the second transition zone. Nonetheless, the second transition is not observed for some of the 2%-loaded nanocomposites. The low volume fraction and large sizes did not help confine the chains, as can be seen in Figure 3.7. As observed for the T_g, we also noted a dependence of the confinement temperature T_{conf} on the nanoparticle size. The constancy of a volume fraction accompanied by the decrease in the nanoparticle size increases the number of nanoparticles and reduces the interparticle distances, which increases the probability of creating confinement sites. The more confinement sites there are in the material, the more confined the chains will be and the higher the T_{conf} value will be (Table 3.1).

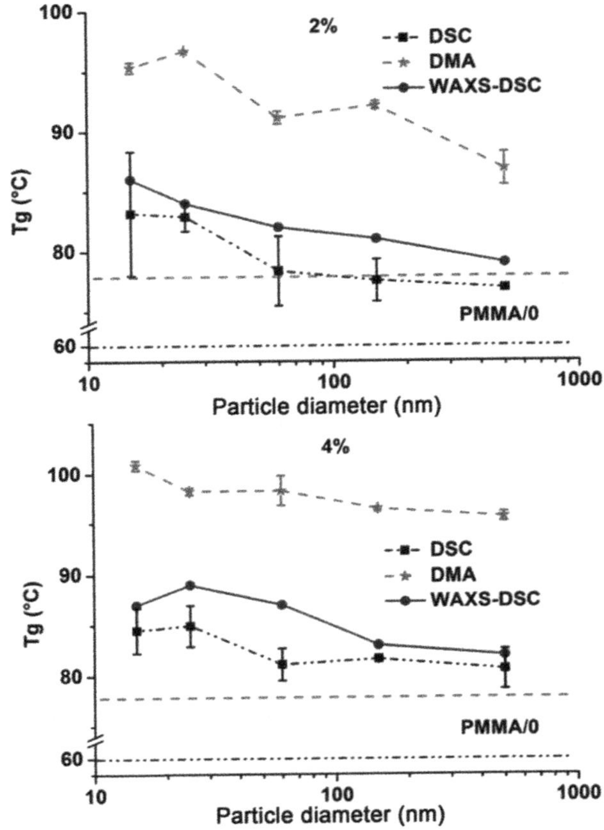

Figure 3.6. *Glass transition temperature of nanocomposites as a function of the nanoparticle size for 2% and 4% volume fractions*

The progressive decrease in the value of d_1 at 30°C can be noted with the decrease in the nanoparticle size. Indeed, its value is 6.9 Å in the case of 500 nm nanoparticles and it decreases progressively until it reaches 6.8 Å for 15 nm nanoparticles (Wang *et al.* 2008). This decrease in d_1 again shows the confinement of the matrix exposed in Chapter 2. In that chapter, the evolution of d_1 was used to show matrix confinement. However, as explained in the same chapter, the evolution of d_1 can also be used to show the same phenomenon.

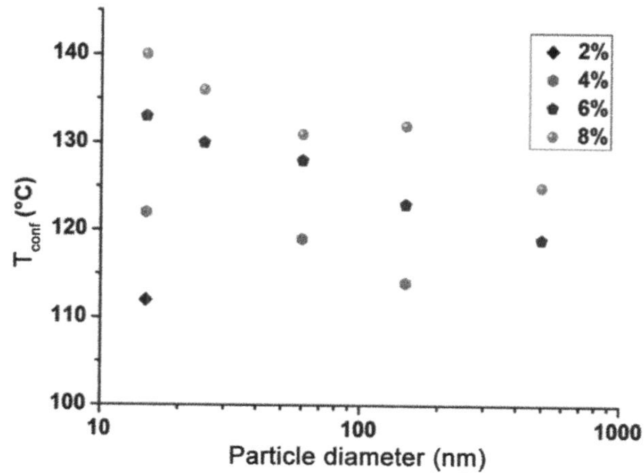

Figure 3.7. *Evolution of the transition temperature of confined domains as a function of the nanoparticle size*

Volume fraction	Size (nm)	T_g (°C) DSC	T_g (°C) DMA	T_g (°C) WAXS	T_{conf} (°C) WAXS
2%	15	83 ± 5	96 ± 1	86	112
	25	83 ± 1	97 ± 0	84	–
	60	78 ± 3	91 ± 1	82	–
	150	77.5 ± 2	92 ± 1	81	–
	500	77 ± 0	87 ± 1	79	–
4%	15	84 ± 2	101 ± 1	87	122
	25	85 ± 2	98 ± 1	89	–
	60	81 ± 2	98 ± 2	87	119
	150	82 ± 1	96 ± 1	83	114
	500	81 ± 2	96 ± 1	82	–
6%	15	83 ± 2	101 ± 1	90	133
	25	84 ± 4	100 ± 3	82	130
	60	85 ± 1	96 ± 1	90	128
	150	84 ± 1	95 ± 0	84	123
	500	82 ± 3	96 ± 0	82	119
8%	15	85 ± 5	100 ± 1	103	140
	25	86 ± 4	100 ± 0	99	136
	60	88 ± 4	98 ± 2	91	131
	150	86 ± 3	97 ± 2	88	132
	500	82 ± 1	95 ± 0	87	125

T_g and uncertainty values are rounded by default or by excess.

Table 3.1. *T_g and T_{conf} values of the different materials*

3.2. Thermal stability

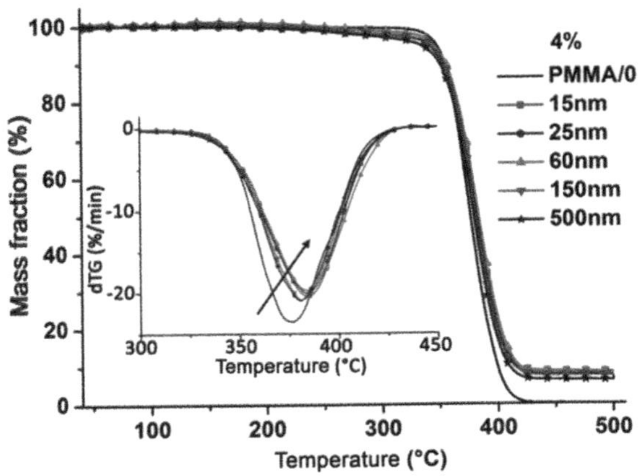

Figure 3.8. *Thermogravimetric analysis and mass loss curves of PMMA/0 and its 4% loaded nanocomposites of different nanoparticles sizes*

The addition of nanofillers to the polymer showed a dependence of T_g on nanoparticle size. The nanoparticle size can also affect the decomposition temperature of the matrix, which governs the polymer thermal stability. The stability study was performed by thermogravimetric analysis under controlled atmosphere (nitrogen). The study consists of studying the different stages of mass loss of the sample (TG) and noting the temperature at which the polymer loses 50% of its mass (dTG). The dTG curve is the derivative of the mass loss curve (TG) and is used for measuring the decomposition kinetics of the polymer. The thermogravimetric analysis curves and mass loss derivatives of PMMA/0 and its nanocomposites filled with 4% of different nanoparticle sizes are shown in Figure 3.8. As for PMMA/0, the thermogravimetric analysis curves of the nanocomposites show a one-step degradation, and this degradation corresponds to the scission of the main chain bonds of the polymer matrix, as shown in the work by Kashiwagi *et al.* (1985). These curves do not help to conclude on a possible effect of presence or size of nanoparticles on the matrix decomposition temperature. However, the derived mass loss curves showed a slight increase in temperature at 50% decomposition (Table 3.2). The temperature for 50% decomposition of the PMMA/0 matrix is 376°C. In the case of 4%/500 nm and 4%/15 nm nanocomposites, the temperatures are 380°C and 385°C, respectively, representing gains of 4°C and 9°C over the matrix. This fact gives a gain of 5°C by decreasing the size from 500 to 15 nm. The size effect, although observed for 4% and 6%, appears to be non-existent at the other volume fractions (2% and 8%). For 2%, a gain of 4°C is observed with respect to the matrix, regardless of the size of the nanoparticles. It is 11°C for 8% fillers for almost all sizes. The improvements in the thermal stability of nanocomposites are due to the fact that the nanoparticles act as a barrier by preventing the residues produced during the reaction from escaping, as observed by Kashiwagi *et al.* (1985).

Size (nm)	2%	4%	6%	8%
15	381 ± 1	385 ± 1	387 ± 1	387 ± 1
25	381 ± 1	384 ± 1	387 ± 1	387 ± 1
60	382 ± 1	384 ± 1	388 ± 1	390 ± 1
150	381 ± 2	381 ± 1	383 ± 1	388 ± 1
500	380 ± 1	380 ± 1	381 ± 1	387 ± 1

Table 3.2. *T50% decomposition temperature values of nanocomposites under nitrogen*

3.3. Effect of size on mechanical properties

Physicochemical characterization has shown that the size of the nanoparticles affects the vitreous transition and decomposition temperatures of the matrix. This effect occurs through the microstructural modification that the variation of the nanoparticle size implies. In the same way that size variation modifies the physicochemical properties, it could also have an effect on the mechanical properties, namely the elastic and viscoelastic properties. These size-dependent properties were measured with adapted tests. The elastic properties were measured by means of instrumented tensile tests and the viscoelastic properties by DMA tests. The testing conditions and results will be presented below.

3.3.1. *Quasi-static tests: elastic properties*

The effect of nanoparticle size on the elastic properties of nanocomposites was determined through an instrumented tensile test campaign following the ISO 527 standard. Dumbbell-like test specimens were utilized with an effective length of 30 mm, a width of 5 nm and a thickness of 1.5 mm. The tests were performed at room temperature (25°C) on a Zwick 4505 uniaxial tensile testing machine with a 500 N force cell (Instron 500 N). To measure the strains, we used KYOWA rosettes, model KFGS-1-120-D16-11 with a strength of 120 Ω, a factor of 2.06, and a gap length of 1 mm. The rosettes are glued and centered on the useful part of the specimens, and are connected to an acquisition chain that transforms the electrical signals into deformations. The rosettes are two-dimensional gauges that allow the measurement of longitudinal and transverse deformations, and thus the determination of the Poisson ratio of materials. The deformation accuracy of the rosettes is 10^{-4}. The imposed speed is 1 mm/min to ensure the quasi-static character of the test. Given that the elastic zone of the polymers is difficult to observe, we used the 0.2% deformation to determine the elastic modulus of our materials. These instrumented tensile tests were performed to determine the influence of the variation of the size of our nanoparticles on the elastic moduli (Young, compressibility and shear) of the nanocomposites at constant volume fraction. The tests were performed for different volume fractions to confirm the repeatability of the size effect. The tests were performed a minimum of three times and the values shown or represented are the averages.

Figure 3.9 shows the stress–strain curves of the PMMA/0 matrix and the 4%/150 nm and 4%/500 nm nanocomposites. PMMA/0 has an average Young's modulus of 3 ± 0.1 GPa, an average flow stress of 57.8 ± 1.4 MPa and an average strain at break of 3.4%. In the elastic region, a stiffening of the material can be noticed with the addition of nanoparticles, and even more so when the nanoparticle size decreases. Similarly, the flow stress increases with the addition of nanoparticles and the decrease in their size. At the same time, a decrease in the strain at break is observed. The Young's moduli of the materials are determined and compared according to the particle size. Based on Poisson's ratio, the compressibility and shear moduli are also calculated and compared according to the size of the particles.

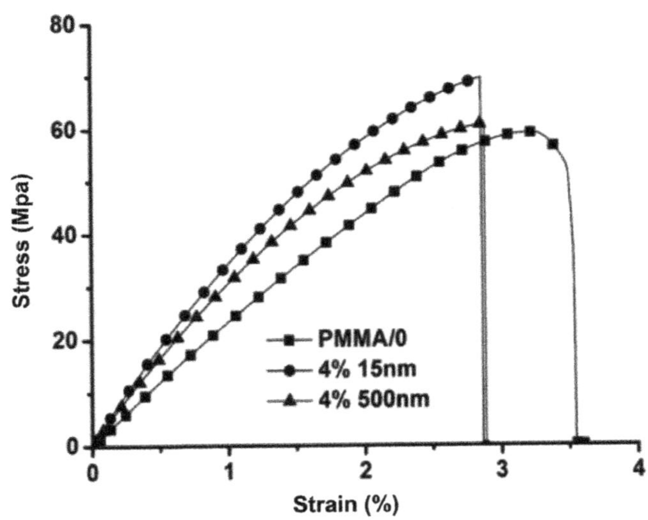

Figure 3.9. *PMMA/0 and 4%/15 nm and 4%/500 nm nanocomposites stress–strain materials response*

The effect of nanoparticle size on the elastic properties of the nanocomposites is clearly shown in Figure 3.9. This figure represents the evolution of Young's moduli, compressibility moduli and shear moduli normalized to the nanoparticle size for the different volume fractions. The moduli are normalized with respect to those of the PMMA/0 matrix. The error bars represent standard deviations from the mean. The values of the Young moduli and the Poisson ratio are recorded in Table 3.3.

The evolutions of the normalized Young's modulus of the nanocomposites are plotted in Figure 3.10 according to the nanoparticle size for the 2%, 4%, 6% and 8% volume fractions. It can be observed that the Young moduli of the nanocomposites are greater than that of the matrix, regardless of the volume fraction and the size of the nanoparticles. These observations result from two combined effects, namely the *effect of the loading rate* and the *size effect*. The effect of the loading rate has been observed and shown experimentally in works in the literature (Usuki *et al.* 1993; Liu *et al.* 2009), while the size effect, although discussed in the literature (mainly in numerical studies), has only been minimally discussed at the experimental level.

	Young's moduli (GPa)			
Size (nm)	2%	4%	6%	8%
15	3.6395 ± 0.1054	4.0428 ± 0.112	3.9479 ± 0.0915	4.4666 ± 0.0671
25	3.58 ± 0.1594	3.8432 ± 0.0758	3.9651 ± 0.0852	4.5268 ± 0.1987
60	3.54 ± 0.0154	3.6686 ± 0.048	3.9203 ± 0.0633	4.4192 ± 0.0951
150	3.48 ± 0.1291	3.6222 ± 0.0906	3.7318 ± 0.062	4.091 ± 0.1841
500	3.39 ± 0.1245	3.5555 ± 0.0623	3.743 ± 0.196	3.8988 ± 0.1102

	Poisson coefficients		
Size (nm)	2%	4%	6%
15	0.3532 ± 0.0011	0.3531 ± 0.0002	0.3535 ± 0.0010
25	0.3640 ± 0.0003	0.3611 ± 0.0060	0.3675 ± 0.0019
60	0.3670 ± 0.0004	0.3696 ± 0.0016	0.3708 ± 0.0015
150	0.3700 ± 0.0015	0.3722 ± 0.0042	0.3732 ± 0.0010
500	0.3740 ± 0.0001	0.3750 ± 0.0033	0.3737 ± 0.0008

Table 3.3. *Values of the Young moduli and Poisson ratio for the different materials*

The observation of the size effect, on the other hand, consists of fixing the loading rate and varying its size. In Figure 3.10, one can also note that the decrease in the size of nanoparticles with a fixed volume fraction is accompanied by an increase in Young's modulus. Indeed, for a volume fraction of 2%, the gain in modulus shifts from 13% with the 500 nm nanoparticles to 21% for the 15 nm nanoparticles, which represents a gain of 9% when decreasing the size from 500 to 15 nm. The same observation is made for the other volume fractions. For example, for 4% volume, the gain in modulus is doubled by decreasing the diameter from 500 (18%) to 15 nm (34%). The size effect is, however, not a linear function of the nanoparticle size. The gain in modulus is small between large nanoparticles and increases very rapidly at small sizes. However, the difficulties related to the dispersion of nanoparticles within the matrix can attenuate the size effect, especially at low sizes where nanoparticles are more reactive with each other. They form aggregates that do not interact in the same way with the matrix as an isolated nanoparticle. In addition, the increase in volume fraction favors the aggregation of small nanoparticles that are more numerous in the matrix. In these cases, the moduli can drop or vary very little, as can be seen for the 6% volume fraction.

By fixing the volume of the nanoparticles that can be found in the matrix and by varying their size, the nanoparticle/matrix contact area is increased, which is related to the nanoparticle diameter. As a matter of fact, at a fixed volume fraction, the decrease in the size of fillers is accompanied by an increase in their total surface area. In the case of nanoparticles, a division of their size by a value α is accompanied by a multiplication of the surface/volume ratio by α. Therefore, the decrease in size increases the total charge transfer surface area between the matrix and the nanoparticles (charge transfer from the soft medium to the rigid medium). The nanocomposites with the smaller nanoparticles thus have a larger charge transfer surface area, which explains the increase in Young's modulus at low nanoparticle sizes. With the increase of their area, small particles have more interaction sites with the matrix, which increases their number of hydrogen bonds. All of these factors improve the adhesion of the nanoparticle to the matrix chains and result in an increase of the Young modulus at low nanoparticle sizes. It is also believed that chain confinement basically modifies the modulus of the matrix. As in the case of thermal properties, where confinement causes a second transition temperature to appear at temperature values above T_g, the same effect can also modify the mechanical properties of the matrix surrounding the nanoparticles and thereby the macroscopic properties. On the other hand, it is difficult to determine the elastic mechanical properties at the scale of the confined phases in order to compare them to the macroscopic mechanical properties, as was the case for the T_g.

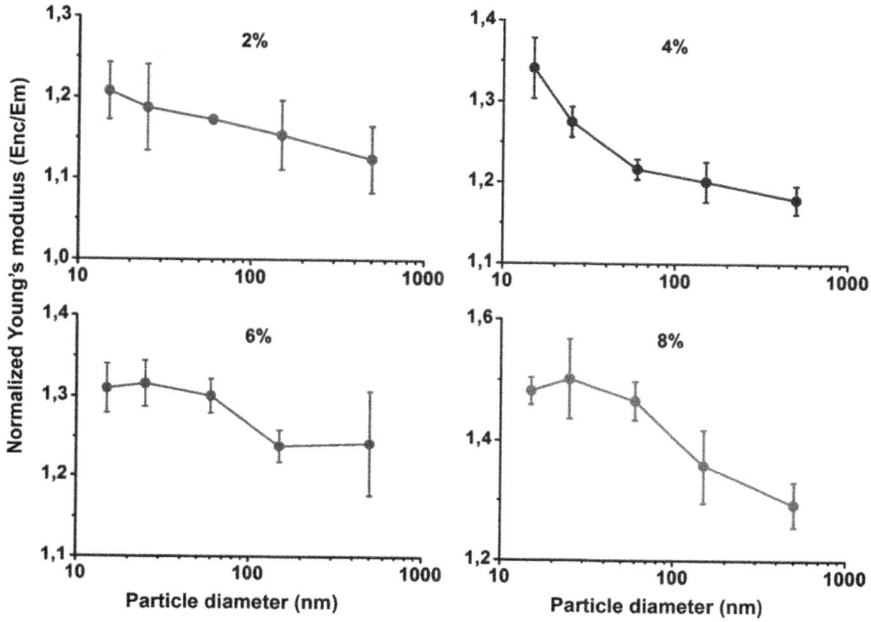

Figure 3.10. *Effect of nanoparticle size on the Young moduli of 2%, 4%, 6% and 8% filled nanocomposites*

By determining the Young modulus and the Poisson ratio, the shear moduli of the materials were calculated and compared for a size effect analysis. The evolution of the Poisson ratio according to nanoparticle size is presented in Table 3.3. It was also noticed that the nanocomposites Poisson's ratios are lower than that of the matrix (Table 3.3) and that they decrease with decreasing nanoparticle size. Given that Poisson's ratio of silica ($v = 0.17$) is lower than that of PMMA ($v = 0.3733$), the addition of the nanoparticles to the matrix therefore leads to slightly decreasing the Poisson ratio, as is the case with the nanocomposites.

Figure 3.11 shows the plot of the shear moduli of the nanocomposites as a function of size for 2%, 4% and 6% volume fractions. The same effect can be observed as that for the Young modulus. The modulus increases slightly at large nanoparticle sizes and then slightly more rapidly at small nanoparticle sizes. The plateau observed at 6% for low nanoparticle sizes is due to a poor dispersion of the nanoparticles in the matrix. The size effect is therefore attenuated.

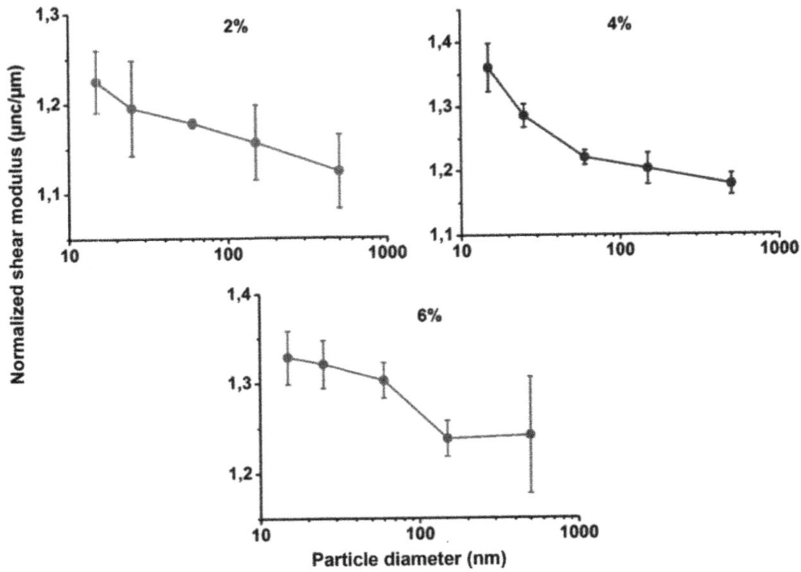

Figure 3.11. *Effect of nanoparticle size on the shear moduli of 2%, 4% and 6% filled nanocomposites*

3.3.2. *Dynamic tests: viscoelastic properties*

Dynamic mechanical analysis was used to measure the conservation modulus (*E*') of the PMMA/0 matrix and the 2%, 4% and 6% filled nanocomposites according to nanoparticle size.

The static tests were done at one temperature and showed that the size of the nanoparticles affects the properties of the nanocomposites. The dynamic tests will make it possible to highlight the effect of size on the viscoelastic properties, and also to determine if the effect of size is effective at high temperatures or inhibited. The tests are done in tension-compression at a frequency of 1 Hz.

Figure 3.12 shows the changes in damping factor (tanδ) according to temperature for PMMA/0 and 4% filled nanocomposites according to nanoparticle size. As the nanocomposite becomes stiffer (increasing *E*'), it loses viscosity (*E*" decreases) and the result is a decrease in tanδ = *E*"/E'. The decrease in nanoparticle size results in a shift of the storage modulus curve toward higher values, which occurs independently of the temperature. As the temperature increases, a more or less rapid drop in the conservation modulus is observed for PMMA/0, while a progressive drop in the modulus is observed below the vitreous transition temperature for the nanocomposites, followed by an abrupt drop above T_g. The presence of nanoparticles in the matrix reduces the rate of fall of the storage modulus and shifts, by the same mechanism, the collapse of the curve toward higher temperatures. It can be noted, for example, that at 30°C, the storage modulus increased from 2.7 GPa for PMMA/0 to 3.17 GPa and 4.19 GPa, respectively, for the 4%/500 nm and 4%/15 nm nanocomposites, which corresponds to respective gains of 17% and 55% compared to the matrix. If we carry out a comparison between the nanocomposites, we obtain a gain of 38% when decreasing the size from 500 to 15 nm for the 4% filled nanocomposites. This improvement of the storage modulus reflects the strong interactions existing between the matrix and the nanoparticles, interactions which are all the stronger as the size of the nanoparticles decreases. It can also be noted that the size effect is maintained until the collapse of the storage modulus curves.

a)

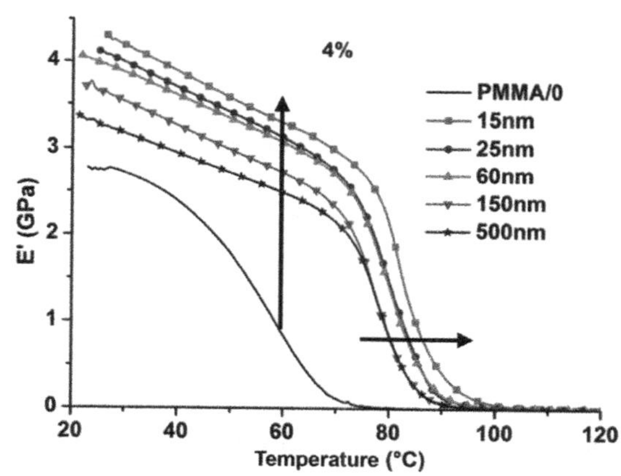

b)

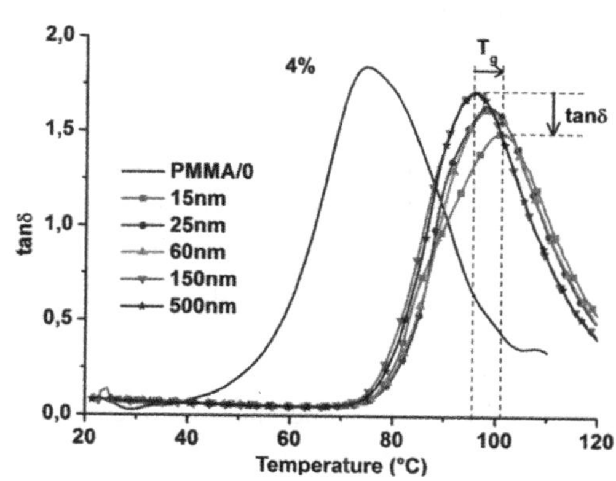

Figure 3.12. *a) Evolution of storage modulus vs. temperature of PMMA/0 and 4% filled nanocomposites with different sizes of nanoparticles and at a frequency of 1 Hz. b) Evolution of the damping factor (tanδ) versus temperature of PMMA/0 and 4% filled nanocomposites with different sized nanoparticles and at a frequency of 1 Hz*

3.4. Conclusion

In this chapter, we have studied the effect of nanoparticle size on the physicochemical and mechanical properties of nanocomposites, and this for different volume fractions.

This study shows, with different techniques, that these properties are dependent on the size of the nanoparticles. An improvement of the properties was observed. A minimal gain of 15°C on the T_g in DSC was obtained when 2% by volume of the 500 nm nanoparticles were added to the matrix, which is a sign of the presence of strong interactions between the nanoparticles and the matrix. This gain increases with decreasing nanoparticle size, it is 21°C at the same volume fraction for 15 nm particles. This improvement was observed for all volume fractions. Another transition zone that was not visible at the macroscopic scale (DSC and DMA) was observed at the molecular scale (WAXS-DSC). This transition zone that appears at higher temperature values than the glass transition zone is evidence of a glass transition gradient induced by the decrease in nanoparticle sizes (Papon *et al.* 2012). The restriction of their mobility is due to their confinement between the nanoparticles, and this confinement becomes more important with the decrease in the nanoparticle size and leads to an increase in the temperature of this transition. This temperature was called the confinement temperature and was denoted by T_{conf}. The value of T_{conf} is also dependent on the size. The confinement state of the chains is not the same from one particle size to another.

As for the mechanical properties, an improvement of the properties was observed with the decrease in the nanoparticle size. Quasi-static tensile tests have shown an increase in Young's modulus with the decrease in the size. This increase is gradual at large nanoparticle sizes and then becomes rapid at small sizes. For 2%, 4% and 8%, the Young modulus increased with decreasing size. In contrast, for 6%, the modulus evolution reached a plateau at small sizes. The size effect is only effective if the nanoparticles are more or less well dispersed. The instrumentation of the tensile test resulted in determining the Poisson ratio and calculating the shear and compressibility moduli. On these two moduli, the size effect was also observed. While it results in an increase in shear, a decrease was observed in compressibility. On the viscoelastic properties, the size effect is summarized by an increase in the storage modulus with the decrease in the nanoparticles size. It also delays the transition zone between glassy and rubbery behavior.

3.5. References

Anoukou, K., Zaïri, F., Naït-Abdelaziz, M., Zaoui, A., Messager, T., Gloaguen, J.M. (2011). On the overall elastic moduli of polymer–clay nanocomposite materials using a self-consistent approach. Part II: Experimental verification. *Compos. Sci. Technol.*, 71, 206–215.

Barnes, J.D., Kolb, R., Barnes, K., Nakatani, B., Hammouda, B. (2000). Microstructure analysis using temperature-induced contrast variation in small-angle scattering. *J. Appl. Crystallogr.*, 33, 758–762.

Blivi, A.S., Benhui, F., Bai, J., Kondo, D., Bédoui, F. (2016). Experimental evidence of size effect in nano-reinforced polymers: Case of silica reinforced PMMA. *Polym. Test.*, 56, 337–343.

Cauvin, L., Kondo, D., Brieu, M., Bhatnagar, N. (2010). Mechanical properties of polypropylene layered silicate nanocomposites: Characterization and micro-macro modelling. *Polym. Test.*, 29, 245–250.

Fornes, T.D. and Paul, D.R. (2003). Modeling properties of nylon 6/clay nanocomposites using composite theories. *Polymer (Guildf)*, 44, 4993–5013.

Gelineau, P. (2015). Caractérisation morphologique et homogénéisation élastique et visco-élastique de polymère. PhD Thesis, Université de technologie de Compiègne.

Kashiwagi, T., Hirata, T., Brown, J.E. (1985). Thermal and oxidative degradation of poly(methyl methacrylate) molecular weight. *Macromolecules*, 18, 131–138.

Kojima, Y., Usuki, A., Kawasumi, M., Okada, A., Fukushima, Y., Kurauchi, T., Kamigaito, O. (1993). Mechanical properties of nylon 6-clay hybrid. *J. Mater. Res.*, 8, 1185–1189.

Liu, Y.Q., Cong, H.T., Wang, W., Sun, C.H., Cheng, H.M. (2009). AlN nanoparticle-reinforced nano-crystalline Al matrix composites: Fabrication and mechanical properties. *Mater. Sci. Eng.*, 505, 151–156.

Maiti, P., Yamada, K., Okamoto, M., Ueda, K., Okamoto, K. (2002). New polylactide/layered silicate nanocomposites: Role of organoclays. *Chem. Mater.*, 14, 4654–4661.

Nguyen, T.L. (2014). Approche multi-échelles dans les matériaux polymères : de la caractérisation nanométrique aux éffets d'échelles. PhD Thesis, Université de technologie de Compiègne.

Painter, P.C., Graf, J.F., Coleman, M.M. (1991). Effect of hydrogen bonding on the enthalpy of mixing and the composition dependence of the glass transition temperature in polymer blends. *Macromolecules*, 24, 5630–5638.

Papon, A., Montes, H., Hanafi, M., Lequeux, F., Guy, L., Saalwächter, K. (2012). Glass-transition temperature gradient in nanocomposites: Evidence from nuclear magnetic resonance and differential scanning calorimetry. *Phys. Rev. Lett.*, 108, 65702.

Paul, D.R. and Robeson, L.M. (2008). Polymer nanotechnology: Nanocomposites. *Polymer (Guildf)*, 49, 3187–3204.

Ryan, A.J., Stanford, J.L., Bras, W., Nye, T.M. (1997). A synchrotron X-ray study of melting and recrystallization in isotactic polypropylene. *Polymer (Guildf)*, 38, 759–768.

Tripp, C.P. and Hair, M.L. (1992). An infrared study of the reaction of octadecyltrichlorosilane with silica. *Langmuir*, 8, 1120–1126.

Urresti, O., González, A., Fernández-Berridi, M.J., Iruin, J.J., Irusta, L. (2011). Oxygen permeability through poly(ethylene-co-vinyl acetate)/clay nanocomposites prepared by micro-wave irradiation. *J. Memb. Sci.*, 373, 173–177.

Usuki, A., Kawasumi, M., Kojima, Y., Okada, A., Kurauchi, T., Kamigaito, O. (1993). Swelling behavior of montmorillonite cation exchanged for ω-amino acids by ϵ-caprolactam. *J. Mater. Res.*, 8, 1174–1178.

Verma, R.K., Velikov, V., Kander, R.G., Marand, H., Chu, B., Hsiao, B. (1996). SAXS studies of lamellar level morphological changes during crystallization and melting in PEEK. *Polymer (Guildf)*, 37, 5357–5365.

Wang, L.F., Pearce, E.M., Kwei, T.K. (1991). Glass transitions in hydrogen-bonded polymer complexes. *J. Polym. Sci. Part B Polym. Phys.*, 29, 619–626.

Wang, W.J., Murthy, N.S., Chae, H.G., Kumar, S. (2008). Structural changes during deformation in carbon nanotube-reinforced polyacrylonitrile fibers. *Polymer (Guildf)*, 49, 2133–2145.

Yang, F. and Nelson, G.L. (2006). Polymer/silica nanocomposites prepared via extrusion. *Polym. Adv. Technol.*, 17, 320–326.

Ying, Z., Xianggao, L., Bin, C., Fei, C., Jing, F. (2015). Highly exfoliated epoxy/clay nanocomposites: Mechanism of exfoliation and thermal/mechanical properties. *Compos. Struct.*, 132, 44–49.

4

Effects of the Size and Nature of Fillers on the Thermal and Mechanical Properties of PEEK Matrix Composites

Marie DOUMENG[1,2], Karl DELBÉ[1], Florentin BERTHET[2], Olivier MARSAN[3], Jean DENAPE[1] and France CHABERT[1]

[1] *ENIT – INPT, University of Toulouse, France*

[2] *Institut Clément Ader, University of Toulouse, France*

[3] *CIRIMAT – ENSIACET – INPT, University of Toulouse, France*

4.1. Introduction

The properties of composites are governed by the properties of each of the constituents and their interactions at the interface. In nanocomposites, this contact surface is larger at constant filler rate, which makes the role of the interface preponderant. This role is even more dominant as the size of the fillers decreases; going from a micrometer to a nanometer filler, the specific surface increases by a factor of 1,000 for spheres. For nanocomposites, the researchers' goal is to predict how matrix chemistry and morphology synergize with surface chemistry, size and shape of nanoscale fillers to define the properties of the resulting material. Crosby (2007) clearly outlines the state of knowledge of the "nano" effect on the mechanical properties of nanocomposites. Since then, numerous works have been able to better quantify the gain brought by nanofillers. This effect comes from the fact that the scales of the macromolecule and the filler are similar. It depends on the matrix, the filler and the interfacial interactions. Crosby and Lee (2007) observe that the "nano"

Nanocomposites,

effect is reduced when nanocomposites are developed by melt-processing, because the nanoparticles are difficult to separate: they form aggregates or coalesce to establish a percolation network within the material. In comparison, when the latter are prepared by *in situ* polymerization and by mixing diluted solutions, the nanoparticles are better dispersed and their effective size is smaller. However, these preparation methods are difficult to transpose industrially. We have therefore chosen to retain the melt method for this work, in order to avoid the use of solvents that are toxic to humans and the environment.

The shape factor (diameter/length, it is equal to 1 for spheres) also plays an important role in maximizing the area/volume ratio. It is clear that the platelet geometry has a significant advantage over more spherical fillers for an equivalent volume fraction. Nevertheless, high form factor fillers are difficult to disperse isotropically at high volume fractions. Consequently, the mechanical properties of the composites are likely to be anisotropic.

The Van der Waals interactions between the particles depend on the chemistry of the filler. The intensity of the interactions is quantified by the Hamaker constant. These interparticle interactions play an important role for high volume fractions by forming a percolating network, and also take part in strengthening particle aggregates.

The size, shape of the particles and their surface chemistry produce the interphase, a zone where the macromolecules are disturbed by the presence of fillers. Within this zone, the mobility of macromolecules, the crystallinity, the entanglement rate and the degree of cross-linking for thermosets are different. Therefore, in the interphase we observe a graduation of the properties, from the interface to the *bulk* matrix. It is commonly accepted that this phenomenon results from the establishment of weak Van der Waals-type bonds and donor–acceptor interactions, as in the case of hydrogen bonds (Jesson and Watts 2012). Numerous studies on polymeric fibers, collected in the review of Liu and Kumar (2014), report that the incorporation of carbon nanotubes enhances the mechanical properties in tension. This is due to the polymer–nanotube interaction that leads to the formation of an interphase, where the polymer is more ordered, more crystalline and better oriented than the rest of the polymer. This interphase is important for the performance of the material, as it cannot only improve the stress transfer between the rest of the polymer matrix and the fillers to fully utilize the mechanical properties of the particles which are stiffer, but also improve the properties of the polymer, as the ordered interphase layer will itself have much better mechanical properties than the rest of the matrix. The same observations were made with functionalized graphene (Terrones *et al.* 2011): the authors found an increase in glass transition temperature of 30°C in polymethyl methacrylate with only 0.05% weight of graphene, and 46°C in poly(acrylonitrile) with 1 wt % of graphene. An increase in the glass transition reflects a mechanical stiffening. These results were attributed to (a) the rough morphology of graphene, which improves the grip with the polymer chain and consequently provides better interfacial adhesion, and (b) the presence of oxygenated groups on the graphene surface, which can form hydrogen bonds with the carbonyl groups of acrylic polymers. Due to the disparity in material properties between the fiber and matrix, stress concentration can develop at the reinforcement–matrix interface. This stress concentration is amplified with the ratio of the shear moduli of the constituents and with the volume fraction of the reinforcements.

Understanding filler/matrix interactions opens the door to modifying the surface of fillers for a better control of mechanical properties. This is demonstrated by Bedi *et al.* (2018), who developed carbon nanotubes on the surface of steel, brass and carbon fibers. The interphase thickness, as measured by scanning electron microscopy (SEM), increases from 7 to 21 μm in the vicinity of steel fibers, for example. The gradual variation of the elastic modulus in the interphase after carbon nanotube grafting suppresses undesirable stress concentrations around the reinforcement caused by the existence of an interface. The review by Karger-Kocsis *et al.* (2015) provides an update on the various avenues explored to modify the surface chemistry of fillers, particularly for carbon fibers. The same idea was exploited earlier by Drzal *et al.* (1983) by way of surface treatment of graphite fibers in an epoxy matrix. The fiber treatment creates a brittle interphase layer between the fiber and the matrix, which increases the interfacial shear strength, resulting in a change in failure mode from interfacial to matrix fracture.

Numerous other studies highlight the effect of nanoscale sizes on the mechanical and thermal properties of composites (Shiu and Tsai 2014; Lee *et al.* 2016; Saleem *et al.* 2016; Khostavan *et al.* 2019). However, it is difficult to establish general rules for correlating these results, as the nature of the interactions in the interphase varies considerably between filler/matrix systems, and the complete chemical definition of their interactions is scarce. In addition, size comparison using multiple types of fillers is rarely the subject of studies. In parallel with experimental studies, modeling efforts have been conducted to better understand the properties of nanocomposites. Among notable works, Choi *et al.* (2015) proposed a model for predicting the overall Young modulus of the material and the mechanical stress transfer to the filler surface as a function of particle size and interphase thickness, using molecular dynamics and finite element simulation.

In this chapter, we have chosen four types of fillers that are known to increase the thermal conductivity of polymer matrix composites and/or provide mechanical reinforcement.

These are alumina (Al_2O_3), silicon carbide (SiC), boron nitride (BN) and graphite. In addition, studies that focus on understanding "nano" phenomena are conducted with model matrices, namely whose rheological behavior is close to the flowing model developed by Flory (1953), and whose resistance to degradation is appropriate for long characterizations. Nevertheless, these matrices are not the most interesting for advanced applications. For this reason, the matrix used is PEEK, a polyaryletherketone (Attwood *et al.* 1981), whose chemical structure made up of aromatic groups linked by ether and ketone groups gives the material a high chemical resistance and mechanical rigidity (Rae *et al.* 2007). PEEK is increasingly used as a replacement for steel in industries that require high-performance materials, such as aerospace or medical ones, with the advantage of considerable weight savings. Some authors have already worked on the influence of mechanical and thermal properties when adding alumina (Goyal *et al.* 2007), carbon nanotubes (Díez-Pascual *et al.* 2012) or graphite (Panda *et al.* 2019). However, the role of the interphase is not detailed in this work. Chen and Hsiao (1992) highlight the appearance of a transcrystallization of PEEK in the presence of carbon, glass and aramid fibers, that is to say, an interphase in the vicinity of the fibers more crystalline than the rest of the matrix. The effect on the thermal and mechanical properties is not specified.

The micro-/nanocomposites prepared by melt-mixing were characterized by thermal measurements, dynamic mechanical analysis and friction measurements to highlight the effect of size and nature of fillers on the thermal and mechanical properties of the composites.

4.2. Materials and methods

4.2.1. *Polymer*

The polymer under study is PEEK 450G, a grade used for injection molding and extrusion, synthesized and marketed by Victrex. This grade is the most common polyetheretherketone used in industry. Its thermal and mechanical properties are given by the supplier and presented in Table 4.1.

Properties	PEEK 450G	Testing method
Melting temperature	343°C	ISO 11357
Start of the glass transition temperature	143°C	ISO 11357
Young's modulus	3.8 GPa	ISO 178
Thermal conductivity	0.32 W/(m·K)	ISO 22007-4
Density	1.30 g/cm^3	ISO 1183
Hardness	84.5 Shore D	ISO 868

Table 4.1. *Properties of Victrex PEEK 450G at room temperature*

4.2.2. *Reinforcements*

4.2.2.1. *Particulate reinforcements*

SiC is a crystalline material formed by the alternating stacking of compact planes of C-Si pairs. More than 200 polytypes of SiC are counted, each of these polytypes has a different structural arrangement. One of them, grade β or SiC-*3C* in Ramsdell notation (Ramsdell 1947), crystallizes in a face-centered cubic lattice whose C-Si stacking sequence contains three layers. Each Si atom binds to four atoms of C forming tetrahedra (see Figure 4.1(a)), which are arranged such that the atoms lie in parallel planes containing nodes of regular hexagons (Lefevre 2008). It has the same structure as diamond carbon. Figure 4,1(b) represents the position of C and Si atoms in the (-1 1 0) plane, A, B and C represent the stacking sequence. In SiC, the interatomic bonds of length 0.189 nm are highly covalent at 89% (Lafon-Placette 2015) due to the small electronegativity difference between Si (1.9) and C (2.55) on the Pauling scale (Pauling 1932). It is used as an abrasive due to its high hardness, but also in the field of electronics for its broadband semiconductor properties. In addition, SiC is chemically inert, it does not react in the presence of a large number of substances and has good corrosion resistance.

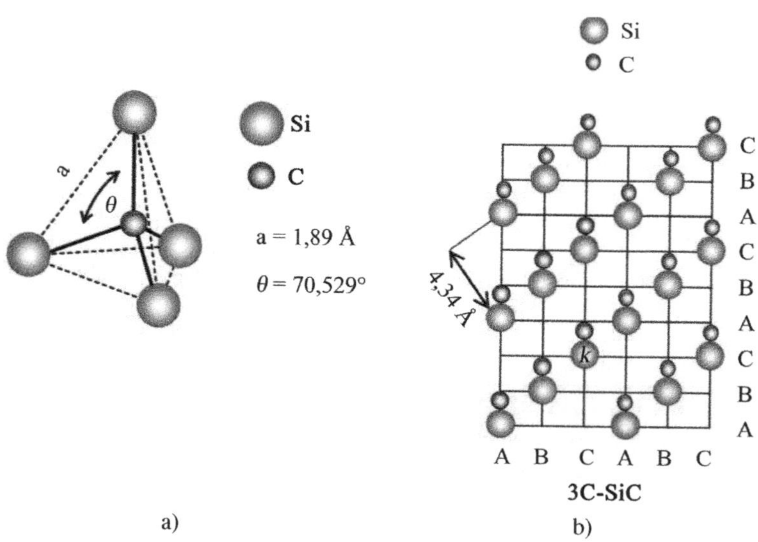

Figure 4.1. *SiC crystal structure: a) tetrahedron formed by four atoms of Si and b) representation in the (–1 1 0) plane of the basic pattern of the polytype 3C (Lefevre 2008). For a color version of this figure, see www.iste.co.uk/bai/nanocomposites.zip*

The α phase of aluminum oxide or alumina, denoted α-Al_2O_3, is a chemical compound existing naturally in bauxite. Alumina crystallizes in the rhombohedral system considered as a compact hexagonal stacking of the O^{-2} oxygen anions with the Al^{3+} cations. The Al atoms occupy two out of three octahedral sites (see Figure 4.2) (Kronberg 1957; Roussel 2013). Alumina contains hybrid ionic bonds at 57% and covalent bonds at 43%, making it resistant to degradation by external agents. Being a ceramic, alumina is chemically inert. Alumina is abrasion resistant, refractory and biocompatible, and could be used in biomedical applications. Its electrical conductivity is low and can be used as an electrical insulator in semiconductor components (Decup 2010).

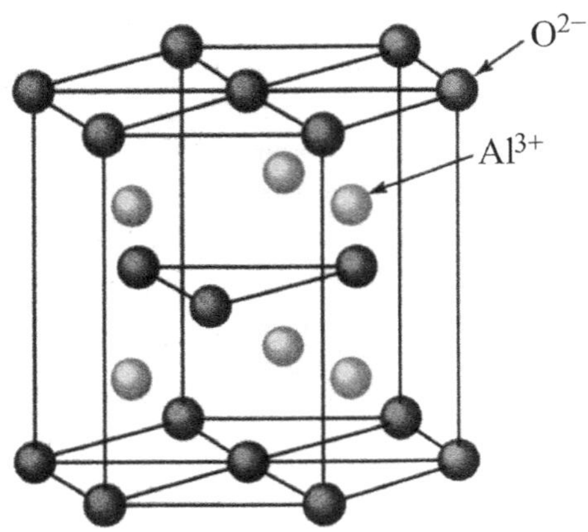

Figure 4.2. *Crystal structure of alumina*

The properties of these reinforcements are summarized in Table 4.2. These particles are expected to reinforce the polymer matrix in volume by stiffening the composite.

Properties	Silicon carbide	Aluminum oxide
Density (g/cm^3)	3.22	3.95
Vickers hardness (HV)	22	13.7
Young's modulus (GPa)	420	400
Thermal conductivity (W/(m·K))	490	30

Table 4.2. *Properties of the particles used in the study at room temperature*

4.2.2.2. *Lamellar compounds*

Graphite is a set of two-dimensional sheets, named graphene, whose thickness is one carbon atom. The sheets are spaced about 0.335 nm apart (Bernal 1924). In natural graphites, two graphite polytypes are mixed: graphite-2H, called α-phase possessing a hexagonal crystal structure, and graphite-3R, known as β-phase possessing a rhombohedral crystal structure (Figure 4.3). This material has lubricating properties, because it is composed of sheets that slide relatively to each other, and the bonds between the sheets are weak Van der Waals-type bonds. In the planes, the bonds are strong and of the covalent type. However, the lubricating properties can be diminished by the presence of gases and vapors in the environment, which limit the interactions facilitating the cleavage of the planes (Savage and Schaefer 1956; Zaidi *et al.* 1995). Thermal and electrical conductivity are high and increase with temperature because phonons and valence electrons have more mobility. Graphite is not chemically inert, it is amphoteric. It can gain or give up electrical charges to form links with its environment. It can therefore form bonds with the polymer matrix.

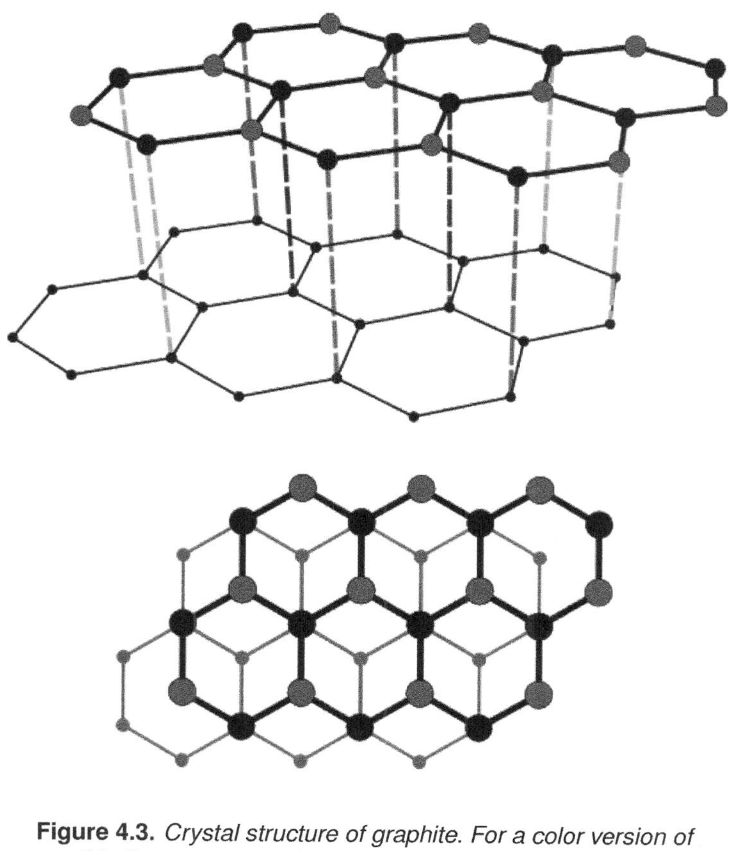

Figure 4.3. *Crystal structure of graphite. For a color version of this figure, see www.iste.co.uk/bai/nanocomposites.zip*

BN is very similar to carbon because it is formed of the elements boron and nitride, which are neighbors of the element carbon in the periodic table. It exists in hexagonal or cubic form. The hexagonal form, denoted h-BN, is a ceramic known as "white graphite" because its color and crystallographic structure is almost identical to that of graphite (Termoss 2009). BN is chemically inert and does not react in the presence of another substance, which makes it different from graphite. Moreover, the shear strain energy dominates over the sliding energy in the case of graphene, which is the opposite for BN (Figure 4.4) (Falin *et al.* 2017). The BN sheets slide more relatively to each other for distances smaller than 15 nm.

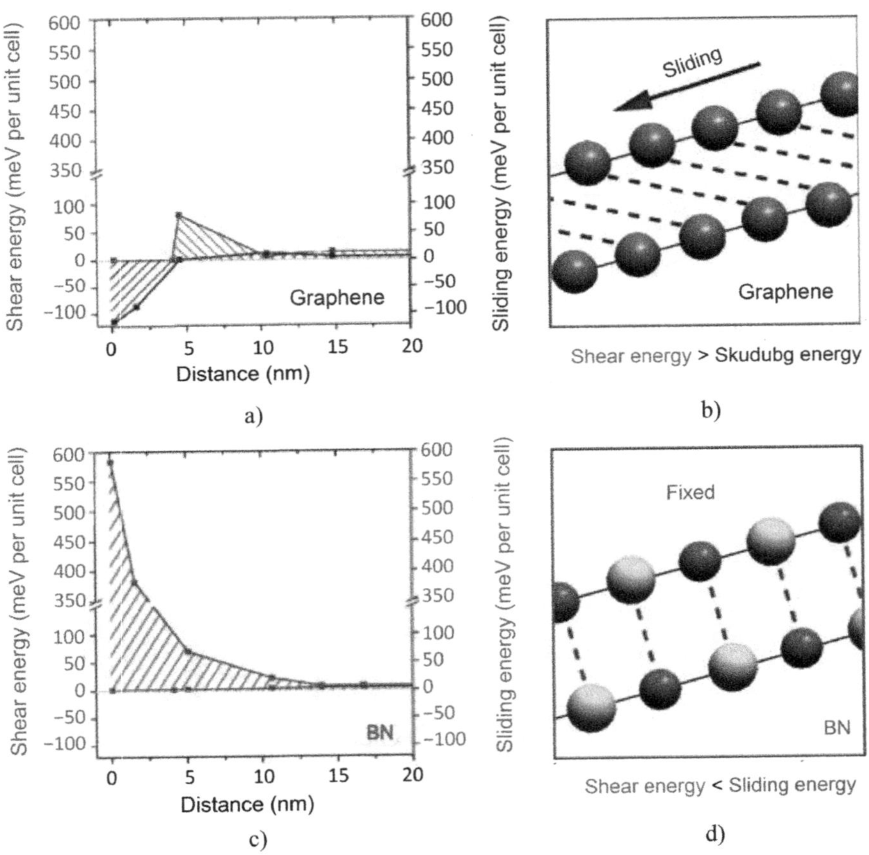

Figure 4.4. *Distribution of shear strain energy and sliding energy according to the distance for a) and b) graphene and c) and d) boron nitride (Falin* et al. *2017). For a color version of this figure, see www.iste.co.uk/bai/nanocomposites.zip*

The properties of these reinforcements are summarized in Table 4.3. The materials are anisotropic, and their properties vary in the plane or perpendicularly to the plane of the sheets. Lamellar compounds are solid lubricants that often limit wear and friction.

Properties	Boron nitride	Graphite
Density (g/cm^3)	2.10	2.10
Vickers hardness (HV)	0.2 ⊥, 24.6 //	60
Young's modulus (GPa)	93	1 000
Thermal conductivity (W/(m·K))	30 ⊥, 600 //	2–800 ⊥, 200–2000 //

Table 4.3. *Properties of the lamellar compounds used in the study at room temperature*

4.2.3. *Nano- and microcomposites preparation*

The composite materials are elaborated by adding powdered reinforcements to the PEEK polymer matrix. To highlight the size effects, two filler sizes are used. The concentrations chosen are from 2.5% to 10% by volume to quantify the effect of the reinforcement rate on the different properties. These rates correspond to those generally used for micro- and nanocomposites. To implement the reinforced PEEK samples, the matrix and reinforcements are blended by melt-mixing with ThermoFischer's Rheomix 600 internal blender, which is equipped with roller rotors suitable for blending molten thermoplastics. The temperature is chosen to be 360°C and corresponds to the complete melting of PEEK on the DSC thermograms. Although the polymer is still very viscous at this temperature, further heating is not possible due to the sensitivity of PEEK to thermo-oxidative degradation. The rotation speed of 40 rpm corresponds to an average shear rate of 40 s^{-1}. When the temperature of 150°C is reached, the mixture is quickly extracted from the chamber before its cooling, which is accompanied by stiffening. The crushed material is placed in a mold, then shaped with the LAB 800P hot press from Pinette Emidecau Industries, requiring only a small amount of material. The compression cycle is composed of three steps. The first consists of heating the polymer to 360°C at 4 K/min to melt it. Then, the temperature is stabilized for 300 s, allowing time for the entire polymer to melt. Finally, the polymer is cooled at a ramp rate of 4 K/min, to allow time for the polymer to crystallize uniformly. When the temperature of the molded plate is below its glass transition, the plate can be removed from the mold without any risk of deformation. To control the roughness, the mold is placed between two sheets of foil, previously cleaned to avoid impurities on the surface of the samples.

4.2.4. *Characterization*

4.2.4.1. *SEM*

The scanning electron microscopic observation is achieved with a ZEISS EVO HD 15 LS to determine the filler distribution in the matrix at a pressure of 40 Pa in the analysis chamber. The observations are made in chemical or topographic contrast mode.

4.2.4.2. *Laser granulometry*

The laser diffraction granulometer instrument is the Horiba LA960 analyzer. It is composed of two light sources and a set of photodiodes collecting the scattered light to calculate the particle size distribution. The analysis is made by a liquid-based process in the aqueous phase with deionized water, in order to respect the integrity of the samples. The ultrasounds are not activated during the measurement. For each sample, two sets of three measurements are performed to check the repeatability and reproducibility of the measurements.

4.2.4.3. *Density*

The density ρ gives an indication of the physical structure of the polymer. The method used in this study is based on hydrostatic weighing or immersion according to the ISO 1183-1:2019 standard. The sample is placed in a nacelle supported on a gantry that rests on the scale. The water beaker is positioned on a base that does not rest on the scale. The sample is cut with an automatic saw to the dimensions 15 mm$\times$ 45 mm. It is then soaked in a wetting agent, a mixture of dishwashing liquid and water, and placed in the weighing device. Each sample is weighed three times.

4.2.4.4. *DSC*

The DSC instrument used is Q200 from TA Instruments. The samples are sliced to obtain about 10 mg of material and then placed in sealed aluminum pans. They are then heated from 80 to 380°C at 10 K/min. A flow of nitrogen is circulated at 50 mm $\cdot$ min^{-1}. The temperature is held for 1 minute, then the sample is cooled from 380 to 80°C at 10 K/min and a heating cycle is applied for the second time. For each material, two samples were analyzed. The DSC analysis gives information on the characteristic temperatures of the polymer, including the glass transition temperature T_g.

The final crystallinity is calculated from the DSC thermogram using the following formula:

$$\chi_c = \frac{\Delta H_m - \Delta H_{cc}}{\Delta H_m^{100\%} \cdot x_M} \cdot 100 \qquad [4.1]$$

The quantities used are the experimental enthalpy of fusion ΔH_m and the theoretical enthalpy of fusion in the case where the material would be 100% crystalline, for PEEK, $\Delta H_m^{100\%} = 130\ \mathrm{J \cdot g^{-1}}$. If cold crystallization occurs, the enthalpy of cold crystallization ΔH_{cc} must be subtracted from the experimental enthalpy of fusion ΔH_m. For a composite, we must take into account its polymer weight content $x_{mas.PEEK}$, which is calculated from the following formulas with the density of PEEK ρ_{PEEK}, the filler density $\rho_{fillers}$ and the filler volume content $x_{vol.Fillers}$:

$$x_{mas.PEEK} = 1 - x_{mas.Fillers} \qquad [4.2]$$

$$x_{mas.Fillers} = \frac{1}{1 + \frac{\rho_{PEEK}}{\rho_{fillers}} \cdot \frac{1 - x_{vol.Fillers}}{x_{vol.Fillers}}} \qquad [4.3]$$

In addition, PEEK can recrystallize when heating during DSC analysis. This phenomenon is not taken into account in the formula of the crystallinity rate. For each sample, two analyses are made to verify the repeatability and reproducibility of the measurements.

4.2.4.5. *Thermal conductivity*

Thermal characterization provides an indication of the material and in particular, its ability to diffuse thermal energy during tribological testing. The Hot Disk TPS 2500S analyzer is an instrument for measuring thermal conductivity, thermal diffusivity and heat capacity of homogeneous and heterogeneous materials in a non-destructive way. The 2 mm diameter probe is made up of a double nickel spiral electrically insulated by a Kapton insulating coating. It provides a heating power fixed at 20 mW for a time of 5 s in order to raise the temperature of the studied material by a few degrees. This same probe is also used to measure the temperature rise by means of recording the variation of its electrical resistance via a Wheastone bridge. The characteristics of the temperature rise, directly related to the evolution of the electrical resistance of the probe, are precisely recorded and the analysis of this variation (transient regime) allows us to determine both the conductivity and the thermal diffusivity. This measurement is absolute and therefore requires no prior calibration or correction factors. Each 3.5 mm thick sample is analyzed five times for test reproducibility.

4.2.4.6. *DMA*

Dynamic mechanical analysis measures the viscoelastic properties and gives information on the molecular mobility of the polymer chains. The tests are performed on a rheometer (ARES LN2, Rheometrics). The configuration chosen in this study is a rectangular torsion with a fixed deformation of 0.5% at 1 Hz imposed by a motor at the bottom axis. The sensor connected to the top axis measures the resulting stress developed in the sample. The sample is heated in an electric oven. The temperature is measured in the air a few millimeters away from the sample. Each material is tested once in the linear viscoelasticity range, that is, the range where stress is proportional to applied strain, from room temperature to 300°C, with a temperature ramp of 3 K/min.

4.2.4.7. *Tribological testing*

The tribological tests are carried out using the UMT Tribolab tribometer from Bruker. All of the tests are performed without lubrication under similar conditions in order to be able to compare them. The chosen configuration is of the ball/plane type, and it leads to having an open contact. The plane is PEEK or PEEK matrix composite. The ball, which is the wiper, is made of 100Cr6 steel with a diameter of 9.5 mm. The parameters are based on standard ASTM G133: "Standard Test Method for Linearly Reciprocating Ball-on-Flat Sliding Wear". The motion is linear reciprocating at a frequency of 5 Hz with a displacement of 10 mm. The normal load of 25 N is controlled by the load sensor, which has a measuring range of 5–500 N with a resolution of 25 mN. Under these conditions, the maximum allowable Hertz pressure for no deformation to occur after withdrawing the ball is 162 MPa, with a contact radius of 0.27 mm for the PEEK sample ($E = 4$ GPa and $\nu = 0.35$). The tests have a duration of 300, 600, 900, 1200 and 1500 s to quantify the evolution of tribological properties over time. Each test is repeated three times to confirm the reproducibility of the tests.

4.3. Results

4.3.1. *Characterization of the powders*

The powders are characterized by SEM and laser granulometry to determine their dimensions before mixing them with the polymer.

4.3.1.1. *Scanning electron microscopy*

Fillers form aggregates and were observed using SEM. The images are presented in Tables 4.4 and 4.5. The dimensions of the fillers are indicated in the tables and are taken from the data sheets provided by the suppliers.

4.3.1.2. *Particle size*

The particle size distribution curves of the initial powders are presented in Figures 4.5(c) and (d), confirming the presence of agglomerates prior to the composite mixing step.

Scale	Silicon carbide	Aluminum oxide
Nano	50 nm (GoodFellow) 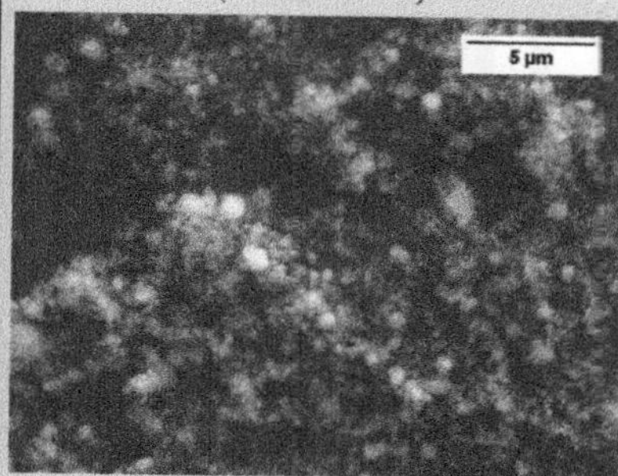	5 to 150 nm (PlasmaChem) 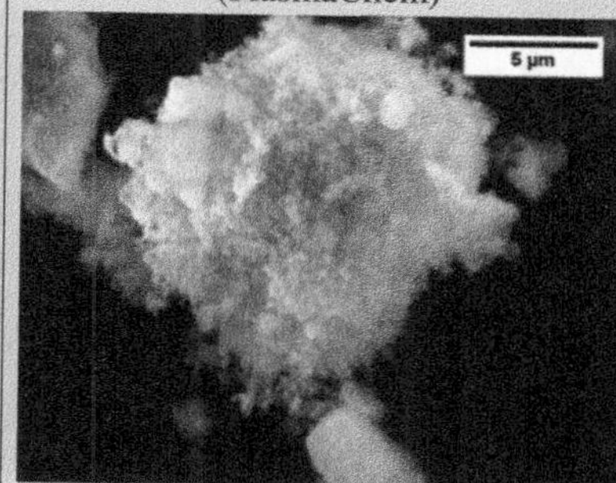
Micro	0.2 to 5 μm (abcr) 	1 to 3 μm (SumitoChem) 

Table 4.4. *SEM observations of the particles used in the study*

For SiC, nano-sized alumina and nano-sized graphite powders, at least two size populations are brought forward. Only the curves for BN powders have a Gaussian and narrower shape, indicating a more regular distribution. SiC is the one whose sizes are the smallest, between 30 nm and 3 μm for nanometric sizes, and between 80 nm and 20 μm for micrometric sizes. Graphite and alumina form the largest agglomerates with maximum sizes of 300 μm. In addition, the powders qualified as nanometric size form agglomerates whose sizes are greater than those of micrometric size powders. It is recalled that these measurements were made without prior dispersion. It is certain that a step in an ultrasonic bath would have resulted in reducing the number and size of the agglomerates. However, and in particular for BN and graphite fillers, the cohesion appears to be weak and the sheets easily detach themselves during handling. For this reason, it is expected that during the mixing step during composite forming, the agglomerates will break up and the laminates will partially exfoliate.

Scale	Boron nitride	Graphite
Nano	0.1 to 1 μm (PlasmaChem) 	3 nm (Avanzare) 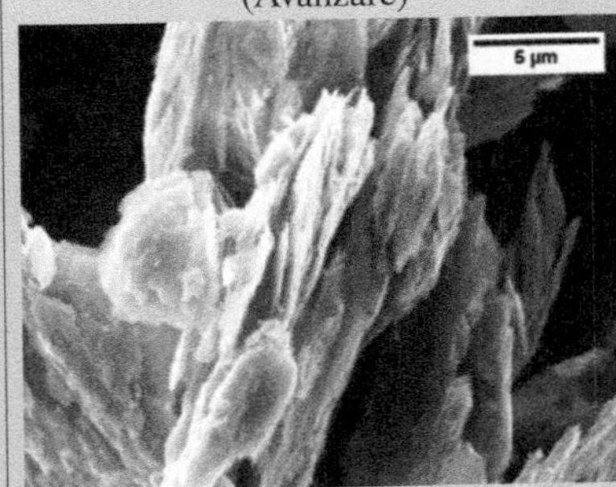
Micro	1 to 10 μm (Saint-Gobain) 	1 to 50 μm (Merck) 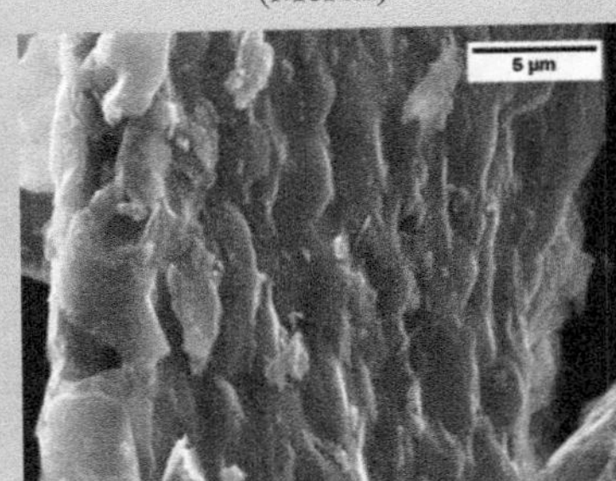

Table 4.5. *SEM observations of the lamellar compounds used in the study*

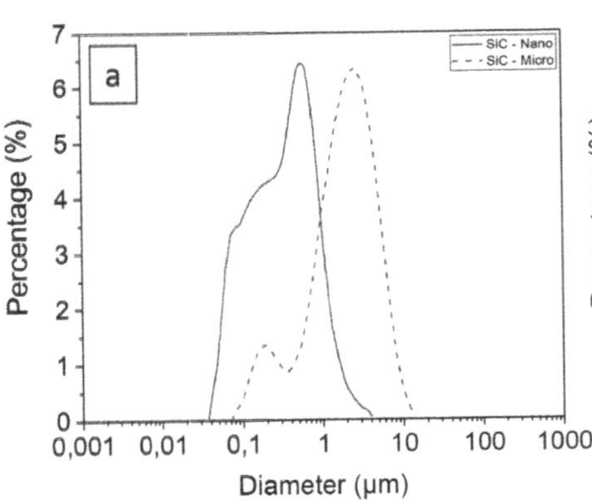

a)

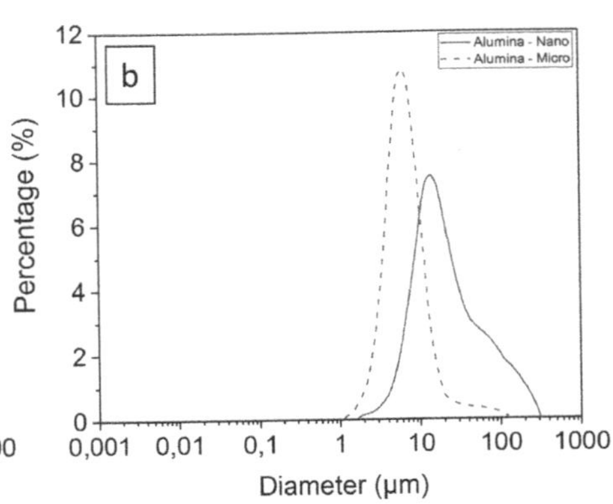

b)

Scale	Silicon carbide	Aluminum oxide
Nano	50 nm (GoodFellow) 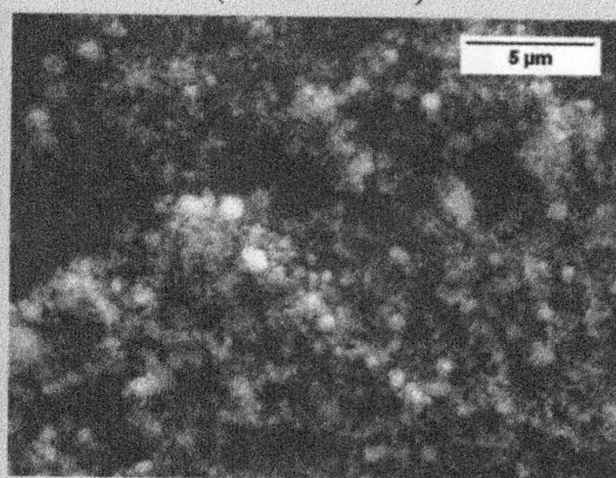	5 to 150 nm (PlasmaChem) 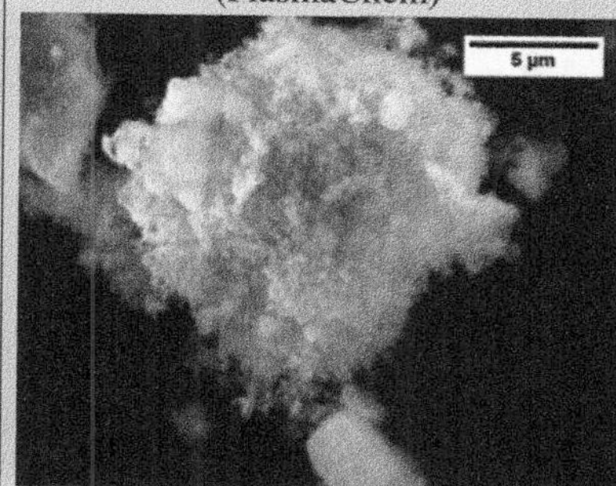
Micro	0.2 to 5 μm (abcr) 	1 to 3 μm (SumitoChem) 

Table 4.4. *SEM observations of the particles used in the study*

For SiC, nano-sized alumina and nano-sized graphite powders, at least two size populations are brought forward. Only the curves for BN powders have a Gaussian and narrower shape, indicating a more regular distribution. SiC is the one whose sizes are the smallest, between 30 nm and 3 μm for nanometric sizes, and between 80 nm and 20 μm for micrometric sizes. Graphite and alumina form the largest agglomerates with maximum sizes of 300 μm. In addition, the powders qualified as nanometric size form agglomerates whose sizes are greater than those of micrometric size powders. It is recalled that these measurements were made without prior dispersion. It is certain that a step in an ultrasonic bath would have resulted in reducing the number and size of the agglomerates. However, and in particular for BN and graphite fillers, the cohesion appears to be weak and the sheets easily detach themselves during handling. For this reason, it is expected that during the mixing step during composite forming, the agglomerates will break up and the laminates will partially exfoliate.

Scale	Boron nitride	Graphite
Nano	0.1 to 1 μm (PlasmaChem) 	3 nm (Avanzare) 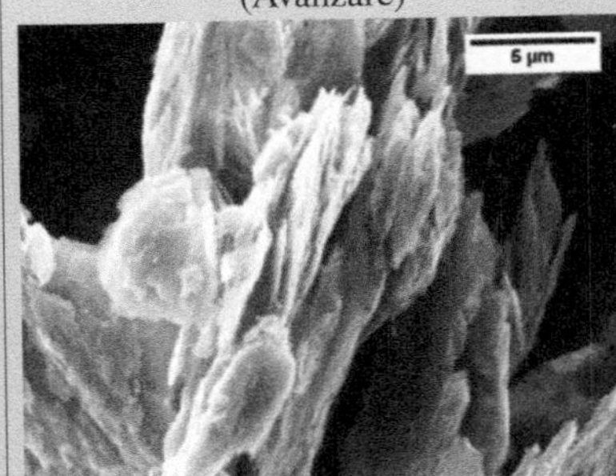
Micro	1 to 10 μm (Saint-Gobain) 	1 to 50 μm (Merck) 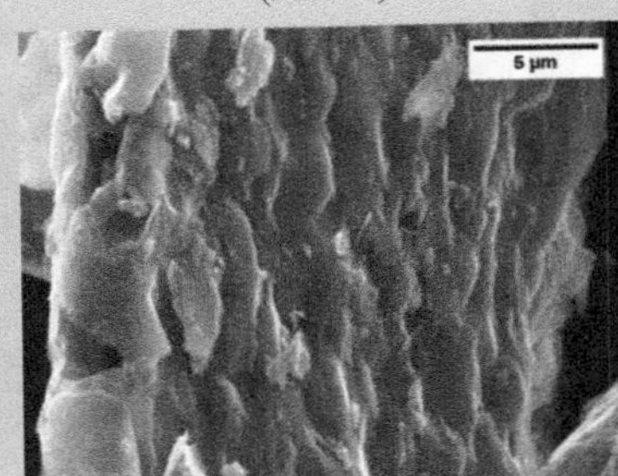

Table 4.5. *SEM observations of the lamellar compounds used in the study*

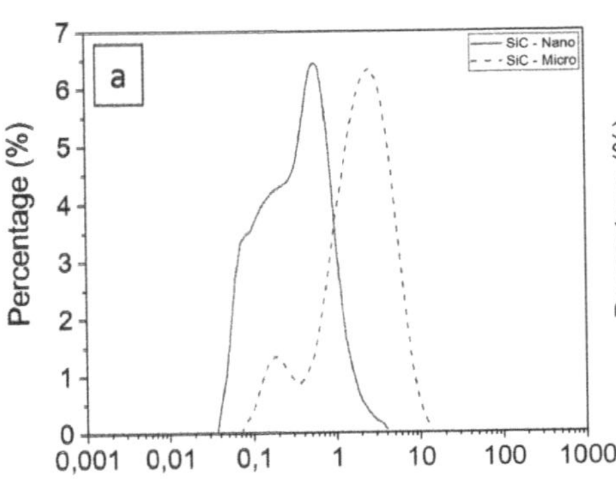

a)

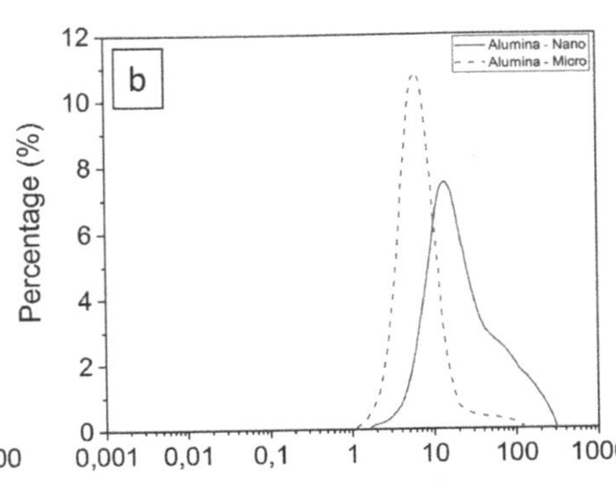

b)

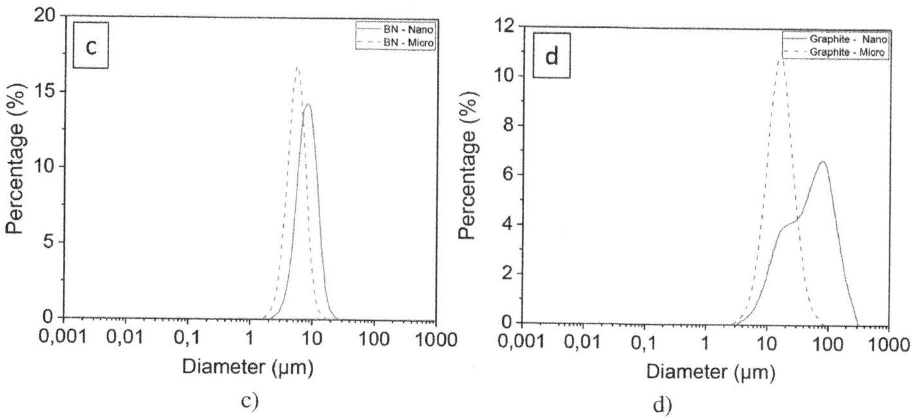

Figure 4.5. *Particle-size curves of nano and micrometer-sized powders of a) silicon carbide, b) alumina, c) boron nitride and d) graphite. For a color version of this figure, see www.iste.co.uk/bai/nanocomposites.zip*

4.3.2. *Filler distribution in the matrix*

The distribution of the fillers has a strong influence on the behavior of the composites, that is why their dispersion in the matrix was verified. As seen previously, all of the fillers used are in the form of aggregates before the mixing step. The high shear viscosity of PEEK in the molten state, around 5000 Pa·s, results in high stress on the aggregates, which leads to filler rupture and erosion. These mechanisms have been well described in the works of Peuvrel–Disdier on polypropylene matrices (Zouari *et al.* 2012; Domenech *et al.* 2014; Normand *et al.* 2016). These works show that the dispersion increases along with the duration of shear until a percolating lattice is created, which is reflected by a drastic increase in viscosity at low shear. Nevertheless, with PEEK, it is not possible to continue mixing beyond 10 minutes due to thermal degradation of the polymer. The dispersion state was quantified by SEM images in chemical contrast mode. The images were analyzed using ImageJ. Figure 4.6 shows the distribution of SiC fillers within the PEEK matrix, SEM images (a and c) and binarized images (b and d). The distribution is considered homogeneous for the two sizes under study.

The histograms illustrate the size distribution of the fillers in Figure 4.7. With the addition of micrometric fillers, particle sizes are larger than those indicated by the suppliers, represented in red on the histograms. This confirms the presence of agglomerates. With the addition of nanoscale fillers, agglomerates are also present, which can reach up to 3 μm. According to the laser particle size results, the D10 value is 80 nm for nanometric SiC; in other words, 10 volume % of the particles are smaller than 80 nm. Nevertheless, the histogram from the image analysis shows that one-third of the detected particles are smaller than 50 nm. These results prove that the mixing step has indeed led to reducing the size of the nanoparticle agglomerates. It is likely that a larger number of even smaller particles are not detected by the analysis software. For micrometer sizes, the value of D50 is at 2 μm: half of the particles measure less than 2 μm before mixing and the image analysis gives about the same proportion after mixing. Therefore, there is little change in size for the micrometer particles following their incorporation into the matrix.

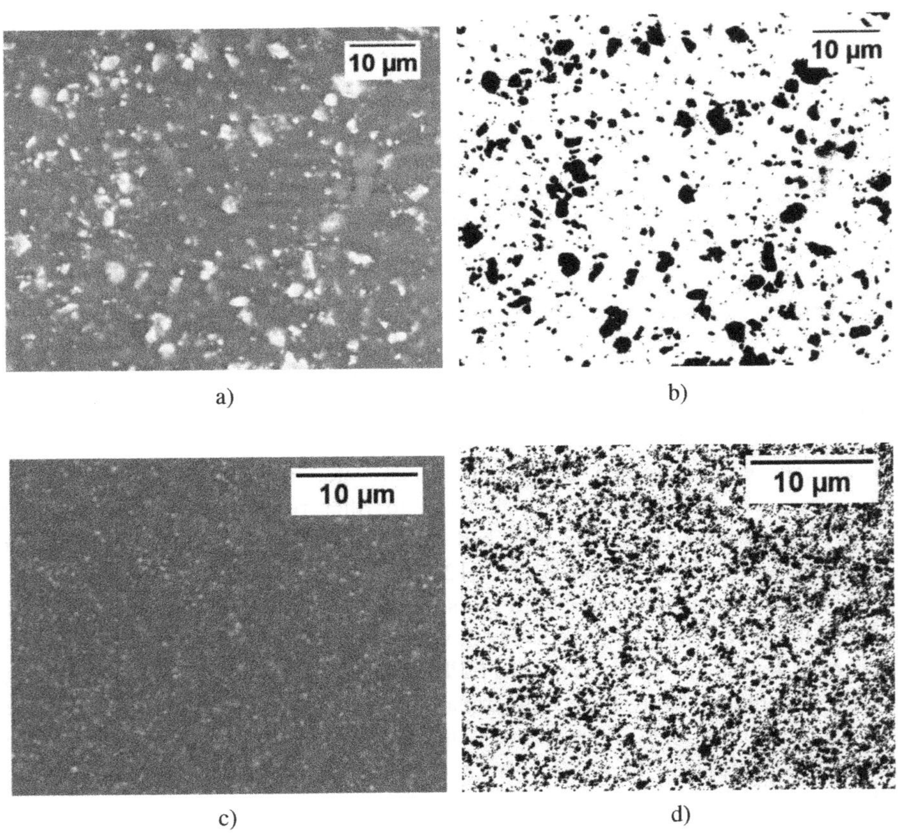

Figure 4.6. *SEM images of the surface of a) nanometer-sized and c) micrometer-sized silicon carbide-loaded samples and associated binarized images b) and d)*

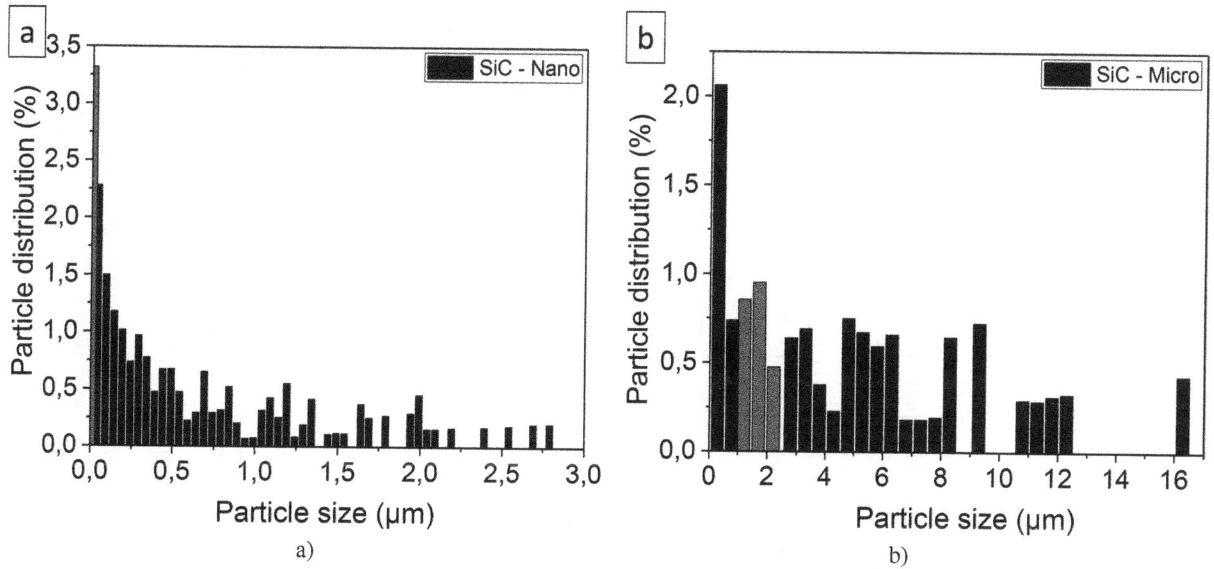

Figure 4.7. *Particle size distribution histograms for silicon carbide loaded samples of a) nanometer and b) micrometer sizes. In red: sizes given by the supplier. For a color version of this figure, see www.iste.co.uk/bai/nanocomposites.zip*

Furthermore, the dispersion is characterized by the critical inter-particle distance *CID*, which is defined by the following equation, where d_r is the diameter of the reinforcements and $x_{vol.Fillers}$ is the content by volume in reinforcements (Kadiyala and Bijwe 2018). It should be noted that this equation is only valid for spherical reinforcements such as alumina and silicon carbide:

$$CID = d_r \cdot [(\frac{\pi}{6 \cdot x_{vol.Fillers}})^{1/3} - 1] \quad [4.4]$$

Figure 4.8 represents the critical interparticle distance for SiC-loaded samples whose shape is similar to a spherical shape. The critical interparticle distances are higher with micrometer-sized SiC particles (CID_{max} = 8.8 μm) compared to nanometer-sized SiC particles (CID_{max} = 0.1 μm). The SiC microparticles are spaced further apart, relatively to each other. The addition of alumina particles gives similar results to the addition of SiC. However, this equation cannot be used in the case of lamellar reinforcements, which do not have a spherical shape.

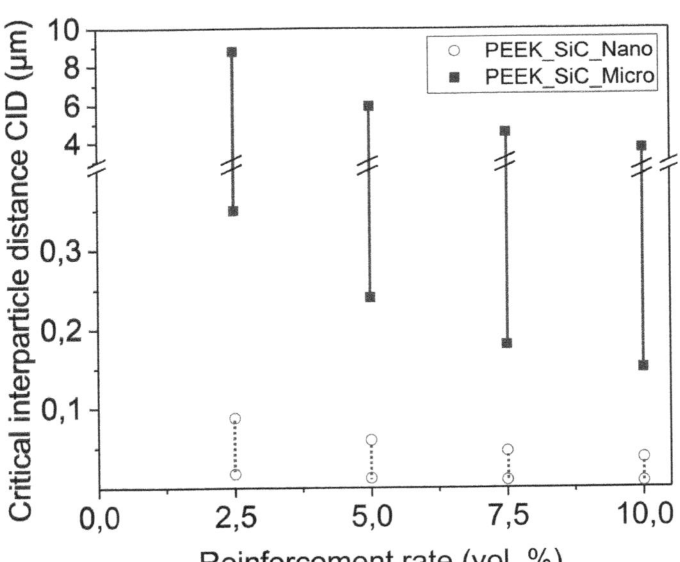

Figure 4.8. *Critical inter-particle distance CID of silicon carbide. The vertical bars indicate the minimum and maximum CID. For a color version of this figure, see www.iste.co.uk/bai/nanocomposites.zip*

In addition, density measurement is performed for all samples to confirm the loading percentage. The density of PEEK ρ_{PEEK} manufactured by thermocompression is 1.297 g/cm^3. The filler densities are: ρ_{BN} = 2.1 g/cm^3, ρ_{SiC} = 3.21 g/cm^3, $\rho_{Graphite}$ = 2.1 g/cm^3 and $\rho_{Al_2O_3}$ = 3.95 g/cm^3. Figure 4.9 represents all of the values of the experimental densities. The density evolves linearly at the reinforcement rates, and the lines representing the theoretical densities according to the rate and type of reinforcement are identical to the theoretical densities. The percentage of fillers corresponds well to the percentage of fillers initially introduced into the polymer.

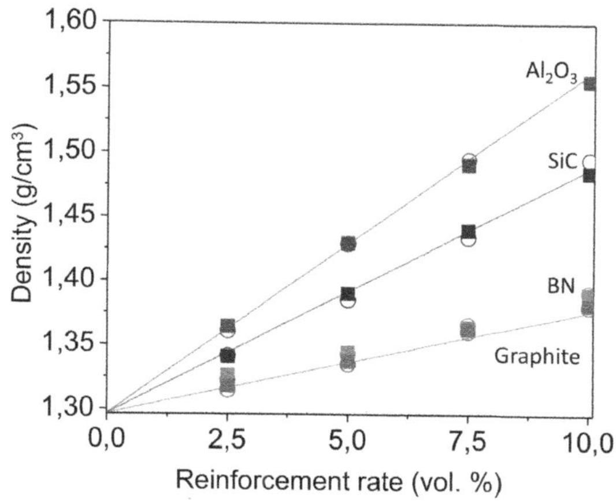

Figure 4.9. *Density of reinforced PEEK samples: nano ○ and micro ■ scale. The absolute uncertainty is ± 0.005 g/cm^3 and is not represented on the plot. For a color version of this figure, see www.iste.co.uk/bai/nanocomposites.zip*

4.3.3. *Effect of size on thermal transitions*

The PEEK thermogram is shown in Figure 4.10. The glass transition temperature is determined by the tangent method by taking the slope at the inflection point. The melting and crystallization temperatures are taken at the maximum of the respective peak. The characteristic temperatures are T_g = 152.5°C, T_m = 345.1°C and T_{hc} = 300.3°C.

An increase in the glass transition temperature T_g means that the polymer material becomes stiffer and retains its mechanical properties at higher temperatures. According to Kadiyala *et al.* (2018), upon addition of reinforcement, the free volume decreases, the critical interparticle distance CID decreases, and the polymer macromolecules have less mobility. However, the effect on the free volume depends on the chemical nature of the filler surface. When the interfacial adhesion is strong, chain mobility is reduced in the interphase area, which results in an increase in the glass transition. Therefore, more energy is needed to allow the macromolecules to move. With the addition of particulate reinforcement in PEEK, T_g decreases, as shown in Figure 4.11. This confirms the existence of a strong interaction between fillers and macromolecules, which leads to the stiffening of the chains. On the contrary, no trend is observed with the addition of lamellar reinforcement with scattered results. This may suggest a heterogeneous dispersion of the reinforcements or the addition of several antagonistic effects on chain mobility. The effect of particles on the melting and crystallization temperature is not shown here. The maximum of the melting peak is shifted by 2°C at most toward the higher temperatures for all of the fillers. Concerning the crystallization phenomenon, the peak maximum is shifted by 8°C at most toward the lower temperatures for SiC and alumina, while it is shifted by 1–2°C toward the higher temperatures for BN and graphite. These changes, even of a small amplitude, confirm the existence of interactions of a different nature in each filler/matrix couple.

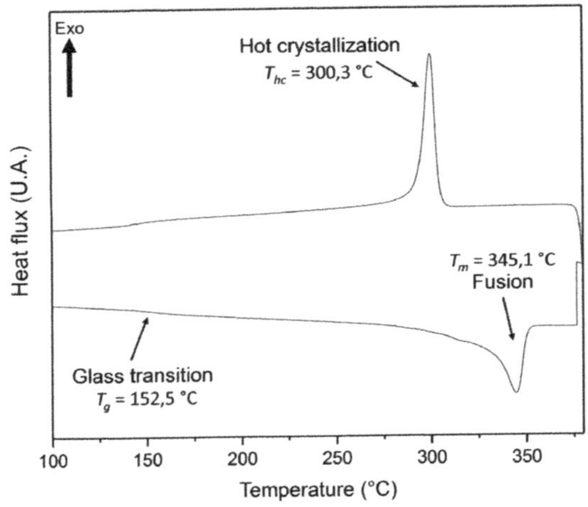

Figure 4.10. *DSC thermogram of PEEK processed by thermocompression*

4.3.4. *Effect of size on the degree of crystallinity*

The degree of crystallinity can be established using several techniques: density measurement, differential scanning calorimetry and X-ray diffraction. We have established an approach to verify the strengths and shortcomings of each technique, in the analysis of the microstructure of PEEK (Doumeng *et al.* 2021). We only use differential scanning calorimetry in this study. The degree of crystallinity is calculated from the enthalpy of fusion and the value is corrected by the actual amount of polymer, taking into account the proportion of fillers. Figures 4.12(c) and (d) illustrate these values for the reinforced samples.

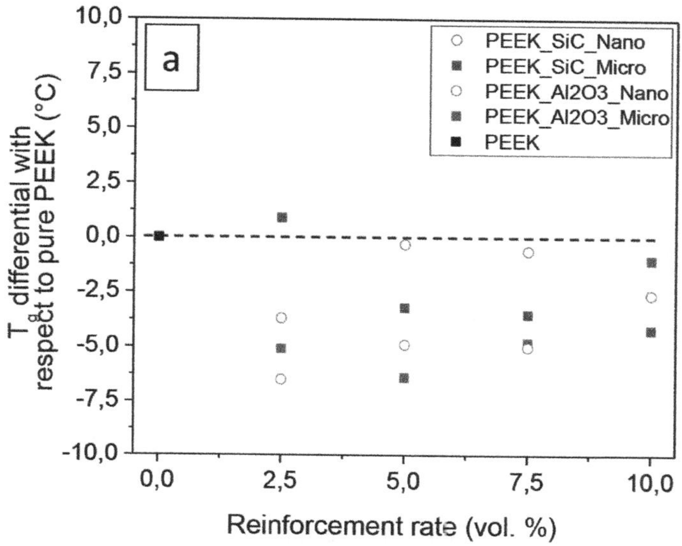

a)

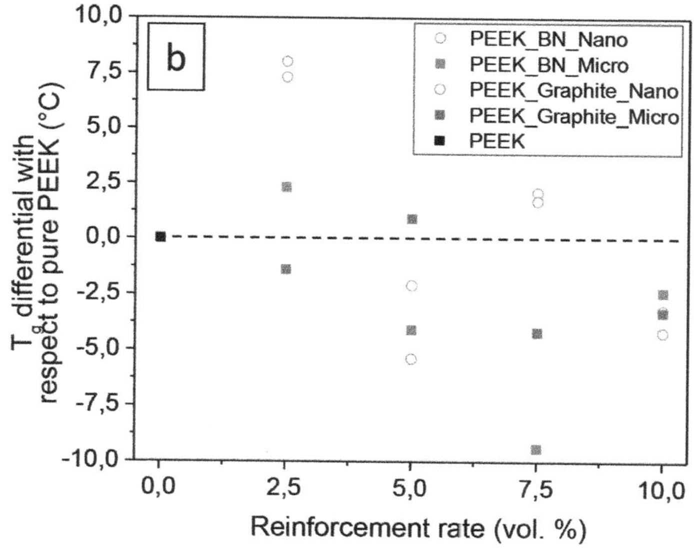

b)

Figure 4.11. *Evolution of glass transition temperatures T_g for PEEK samples reinforced with fillers: a) particulate and b) lamellar. The absolute uncertainty is $\pm$ 0.1°C, and it is not shown in the figure. For a color version of this figure, see www.iste.co.uk/bai/ nanocomposites.zip*

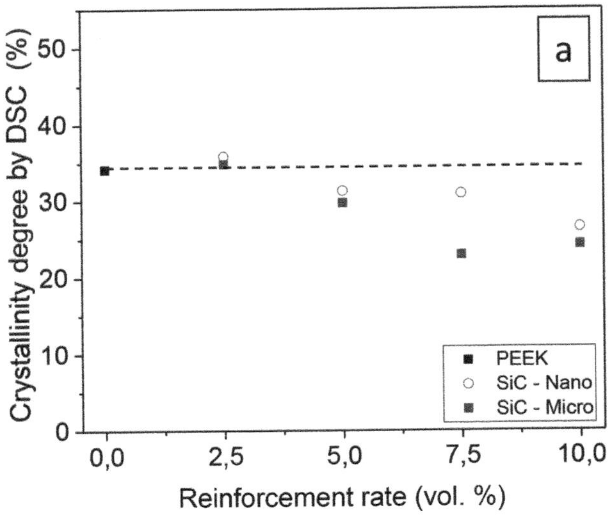
a
Crystallinity degree by DSC (%)
Reinforcement rate (vol. %)
PEEK
SiC - Nano
SiC - Micro
0
10
20
30
40
50
0,0
2,5
5,0
7,5
10,0

a)

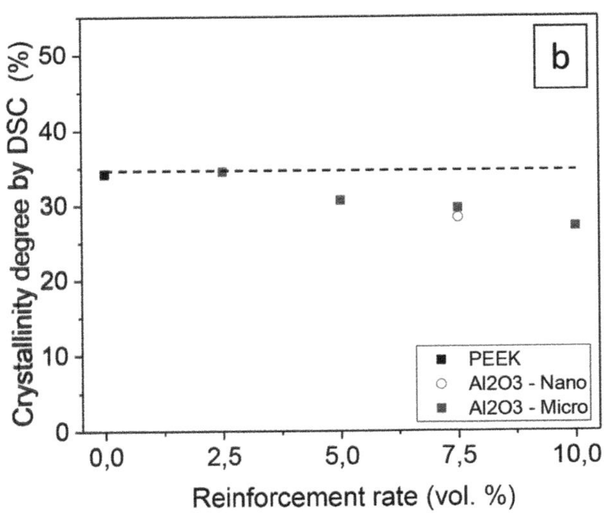
b
Crystallinity degree by DSC (%)
Reinforcement rate (vol. %)
PEEK
Al2O3 - Nano
Al2O3 - Micro
0
10
20
30
40
50
0,0
2,5
5,0
7,5
10,0

b)

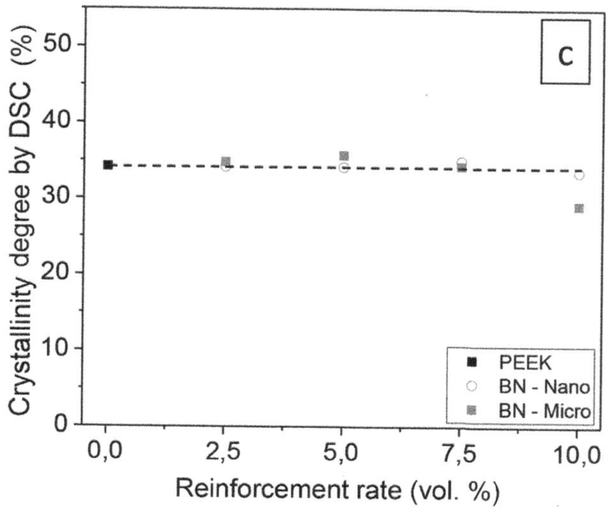

c)

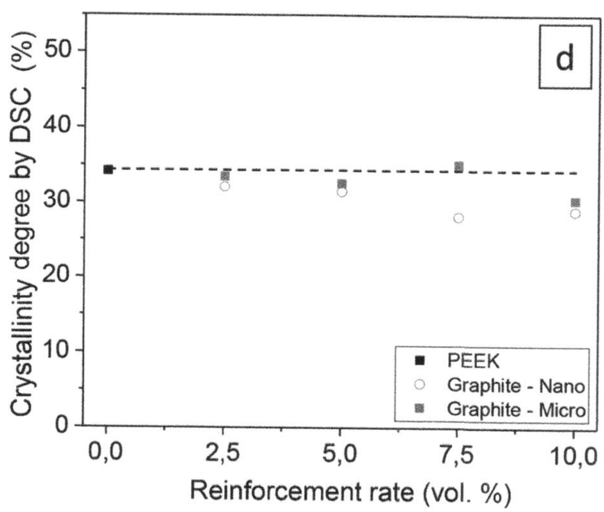

d)

Figure 4.12. *Crystallinity degree of reinforced PEEK samples determined by DSC: nano ◯ and micro ■ scale. The absolute uncertainty is ± 0.4%, and it is not shown in the figure. For a color version of this figure, see www.iste.co.uk/bai/nanocomposites.zip*

Particulate-based reinforcements (alumina and SiC) imply a greater decrease in the crystallinity degree compared to lamellar-based reinforcements (BN and graphite). These results are in agreement with the findings on the evolution of the crystallization temperature. The addition of SiC and Al_2O_3 results in a hindrance for the macromolecules, they have to self-organize. This hindrance could originate from a stronger interfacial adhesion. The movement of the chains would then be limited by their attachment to the particle. Kuo *et al.* (2005) observed a decrease in the crystallinity degree from 5.5 weight % (that is, 1.9 volume %) of alumina, whereas Kadiyala *et al.* (2018) observed a decrease from 10 weight % (that is, 4.3 volume %) of SiC. The size of the reinforcements does not seem to significantly influence the degree of crystallinity. For all of the samples, there still seems to be a tendency for the crystallinity to decrease with the reinforcement rate.

4.3.5. Thermal properties

Thermal conductivity, thermal diffusivity and mass heat capacity were measured for PEEK and for each of the composite materials in this study.

We recall that the thermal conductivity represents the ability of the material to gradually transport thermal energy. Thermal diffusivity represents the capacity to distribute more or less quickly and homogeneously this thermal energy in the material; it can thus depend strongly on the homogeneity of the material. Finally, the mass heat capacity is associated with the thermal energy to be brought to the material to raise its temperature by 1 kelvin per unit of mass. It also reflects the degree of freedom of the elementary constituents of the material: therefore, the higher the heat capacity, the more the phenomena of vibration, rotation and deformation at the molecular scale are numerous and can be brought into play. These three thermal properties are related since the thermal conductivity is the product of the density with the thermal diffusivity and the mass heat capacity.

The thermal conductivity of the fillers is higher than that of PEEK, which is 0.29 W/(m·K). Therefore, the composites have a higher conductivity than the virgin matrix. This is well highlighted in Figures 4.13(a) and 4.14(a).

With materials containing SiC or Al_2O_3, the thermal conductivity increases with filler content from 0.3 to 0.5 W/(m·K) (see Figure 4.13(a)). The particle size does not seem to significantly affect this increase. At the same time, the thermal diffusivity of the composites increases significantly: from $2.5 \cdot 10^7$ to $3 \cdot 10^7$ m^2/s (see Figures 4.13(b) and 4.14(b)). The addition of manometric SiC leads to the most notable increase in thermal diffusivity. Thereby, SiC nanoparticles contribute to the most efficient heat diffusion of this series of materials. Finally, the heat capacity generally varies little with the addition of SiC or Al_2O_3 filler (see Figure 4.13(c)). Nonetheless, the SiC nanoparticles lead to a significant lowering of heat capacity, which is consistent with a decrease in intramolecular mobility that would be caused by the presence of SiC nanoparticles. The 2.5 volume % content of micrometric SiC also causes this drop in heat capacity, but this effect quickly fades with the addition of fillers.

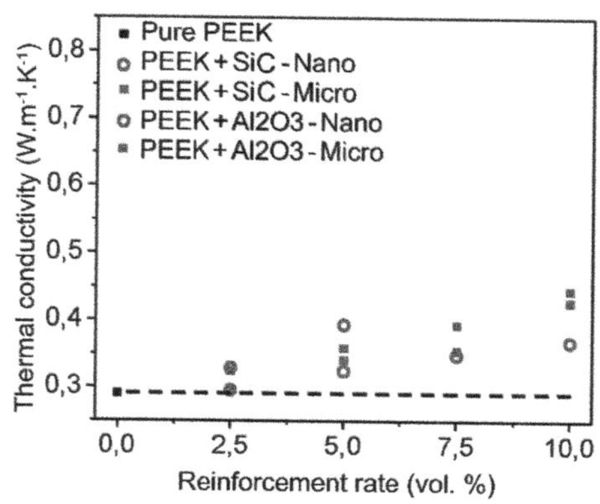

a)

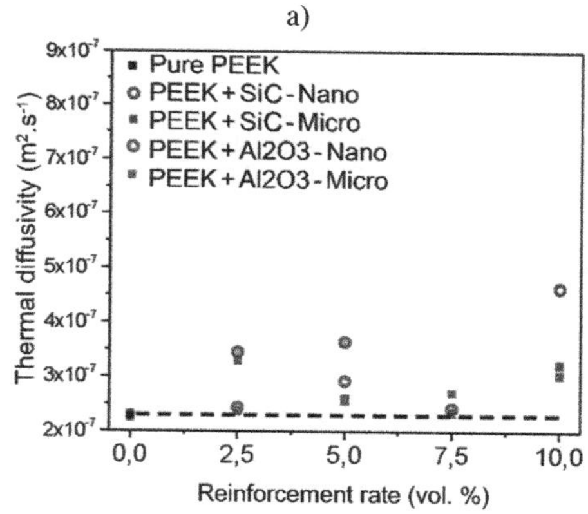

b)

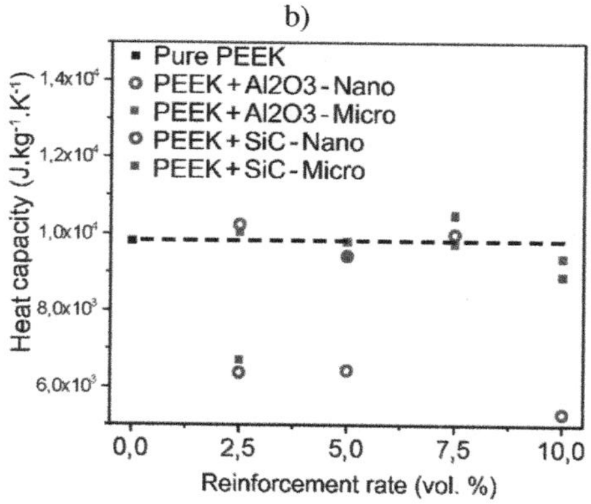

c)

Figure 4.13. *Evolution of a) thermal conductivity, b) thermal diffusivity and c) heat capacity of particulate filler-reinforced PEEK samples. Nano ○ and micro ■ scale. For a color version of this figure, see www.iste.co.uk/bai/nanocomposites.zip*

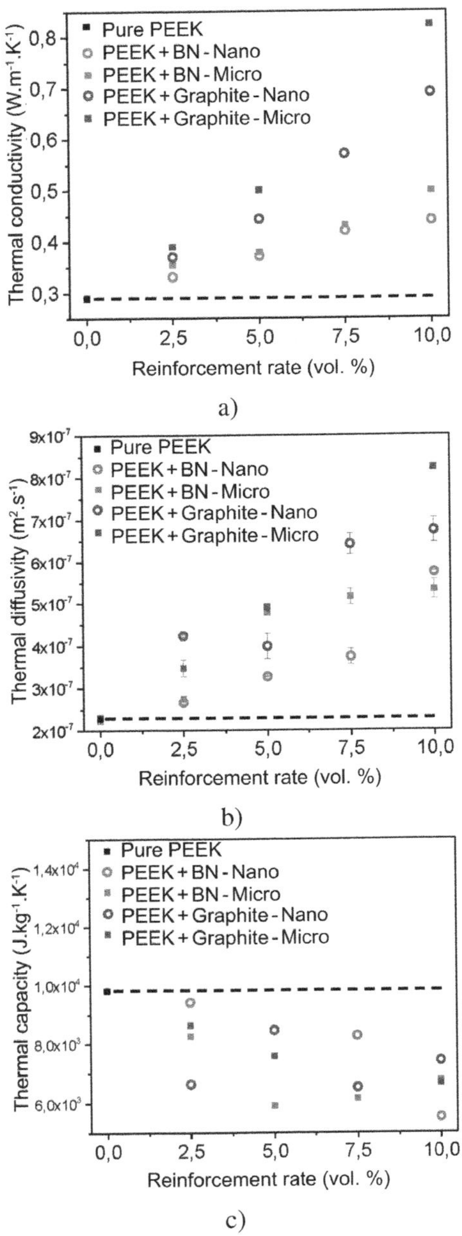

Figure 4.14. *Evolution of: a) thermal conductivity, b) thermal diffusivity and c) heat capacity of lamellar filler-reinforced PEEK samples. Nano ○ and micro ■ scale. For a color version of this figure, see www.iste.co.uk/bai/nanocomposites.zip*

In the case of lamellar particles, the increase in thermal conductivity is more significant, from 0.35 W/(m·K) with 2.5 volume % of lamellar fillers, up to 0.4 W/(m·K) with 10 volume % of h-BN. We even measure 0.8 W/(m·K) with 10 volume % of micrometric graphite (see Figure 4.14(a)). This is accompanied by an increasing thermal diffusivity with particle content (see Figure 4.14(b)) and a decrease in mass heat capacity (see Figure 4.14(c)). The addition of lamellar particles increases the ability of the material to transport heat efficiently and homogeneously throughout the volume; however, this is at the expense of intramolecular mobility. It is therefore the networks created by the additional fillers that seem to cause the improvement of the thermal conductivity of these materials.

PEEK loaded with 7.5 volume % of SiC nanoparticles or 5 volume % of graphite nanoparticles does not seem to have a homogeneous filler distribution in the matrix. As a matter of fact, the thermal diffusivity and heat capacity of these two materials do not follow the generally observed trends.

4.3.6. *Effect of size on mechanical properties*

4.3.6.1. *Thermomechanical study*

Although the long-range molecular motion of PEEK is reduced below the α transition corresponding to the glass transition temperature, there exists limited motion of the lateral groups, corresponding to the β and γ transitions. The β transition takes place at -95°C, reflects the rotation of aromatic rings and movements of polar carbonyl bridges, and is more pronounced in the presence of moisture. A γ transition has also been reported at a temperature of -155°C, resulting from very localized non-cooperative motions (David and Etienne 1992).

Our tests are conducted between 50°C and 300°C; within this range, only the α transition is visible. The curves showing the evolution of the storage modulus E', the loss modulus E'' and the ratio $\tan\delta$ are presented in Figure 4.15. The storage modulus of PEEK is stable up to 145°C with a value close to 1 GPa. The polymer is found in its glassy state in which it is rigid and brittle, due to the reduced molecular movements. Between 145°C and 185°C, a drop of one decade in modulus is observed, which corresponds to the shift to the glass transition until about 200 MPa is reached. From 185°C, E' decreases more slowly, it is the rubbery plateau in which molecular movements are possible, the material is soft. From 270°C onwards, the fall of the modulus accelerates, it is the beginning of the flow of the polymer when the molecules begin to slide on one another. The loss modulus E'' is equal to 10 MPa, before the glass transition. In this glassy state, E' is greater than E''; this is characteristic of a "solid" behavior of the viscoelastic material, which is preserved throughout the temperature range studied. On the contrary, above 300°C (not shown), E'' becomes dominant, the material then exhibits a "liquid"-like behavior. In the transition region between 125°C and 225°C in Figure 4.15, the value of E'' increases sharply and then decreases; this peak indicates the maximum amount of heat dissipated per deformation unit. At the molecular level, this maximum value of the loss modulus indicates that intermolecular bonds are broken, macromolecules have more mobility to flow by rubbing against each other. The *tan* δ curve follows a shape identical to that of E'', with a maximum at the time of the glass transition.

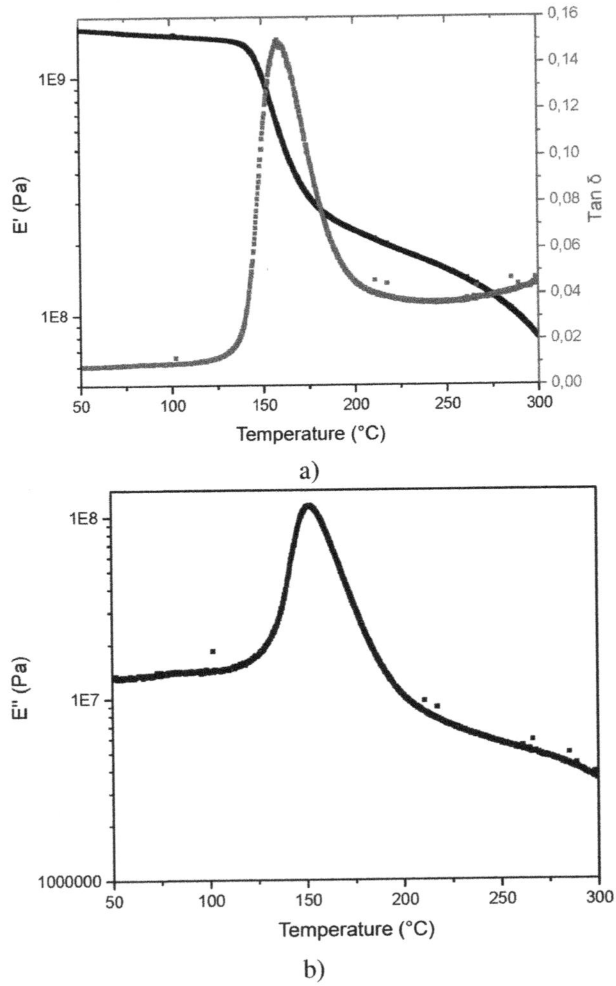

Figure 4.15. *Evolution of a) the storage modulus E' of the loss angle tan δ and b) the loss modulus E'' for PEEK*

The storage modulus E', corresponding to the proportion of energy recovered, increases with the addition of reinforcement for all samples at room temperature; thereby, the energy return is greater. To highlight this evolution, Figures 4.16(c) and (d) represent the evolution of E' at 50°C according to type, size and loading rate. In our study, the evolution of E' is linear according to the content for this concentration range. However, the phenomenon is more complex, E' is the sum of the contribution of several phenomena: filler properties, interfacial interaction and organization of the fillers within the matrix. We assume that the evolution of E' follows a mixing rule in our particular case; we have drawn lines on the graphs to quantify this evolution.

The increase in E' is more significant with the addition of micron BN reinforcement with a slope of $9.4 \cdot 10^7$ Pa/% for the fitting curve. Nevertheless, this is a fitting curve that follows a mixing rule and does not take into account interactions between the reinforcements and the matrix. On the contrary, the smallest increase in E' corresponds to the nanoscale alumina reinforcements, with a slope of $2.8 \cdot 10^7$ Pa/% for the fitting curve. This means that the proportion of energy recovered is greater with micrometric BN reinforcements, and the material is stiffer than with nanometric alumina reinforcements. We note that the sample reinforced with 2.5 volume % of boron nitride microfiller has a storage modulus E' of 1.4 GPa, which is lower than that of PEEK. The energy recovered is less important for this sample. Nonetheless, it appears that the points corresponding to the concentrations under study are aligned. This shows that the increase in stiffness of PEEK with BN is not an affine equation passing through 0 volume %. It appears that BN causes a decrease in the stiffness of PEEK, which is explained by the creation of an interphase of greater mobility around the BN layers, due to the absence of interfacial adhesion.

Let us now look at the loss modulus, which reflects the viscous character of the material. It represents the flow of chains, one on top of the other, without modification of their chemical structure. Due to molecular friction, heat is dissipated; this energy is lost by the material. When E'' is higher, it means that the proportion of energy dissipated will be higher. As mentioned before, the evolution of E'' is linear in our case, but other phenomena are taken into account, such as the filler distribution within the matrix and the interactions between fillers and matrix. If we compare our samples, E'' is the highest for the graphite reinforced sample (see Figures 4.17(c) and (d)). For particulate filler-reinforced samples, the values of E'' vary little with the addition of reinforcement.

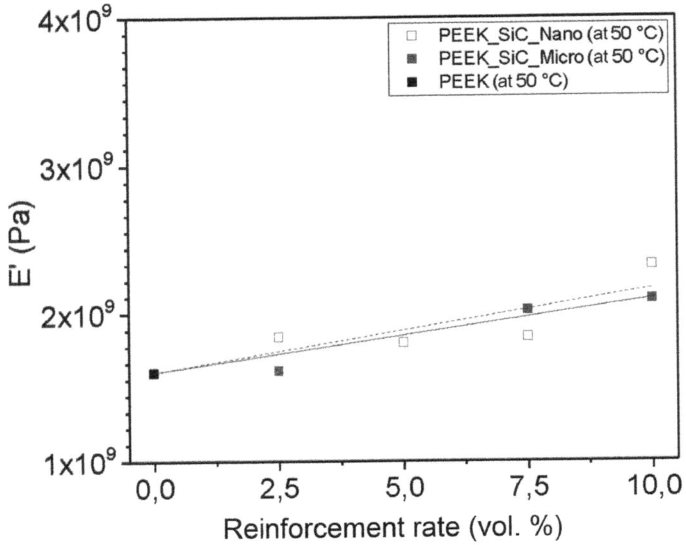
PEEK_SiC_Nano (at 50 °C)
PEEK_SiC_Micro (at 50 °C)
PEEK (at 50 °C)
E' (Pa)
4x10⁹
3x10⁹
2x10⁹
1x10⁹
0,0
2,5
5,0
7,5
10,0
Reinforcement rate (vol. %)

a)

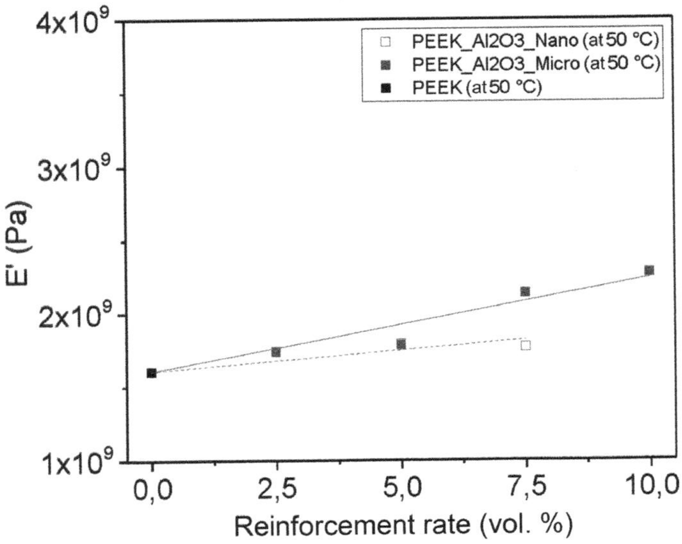
PEEK_Al2O3_Nano (at 50 °C)
PEEK_Al2O3_Micro (at 50 °C)
PEEK (at 50 °C)
E' (Pa)
4x10⁹
3x10⁹
2x10⁹
1x10⁹
0,0
2,5
5,0
7,5
10,0
Reinforcement rate (vol. %)

b)

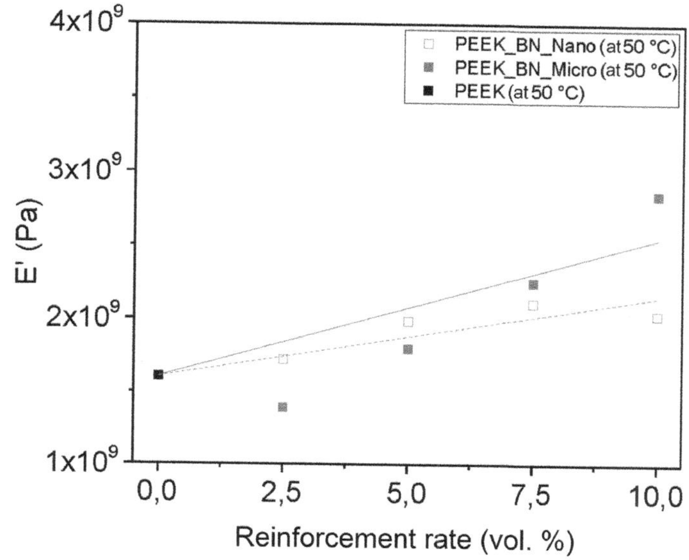

c)

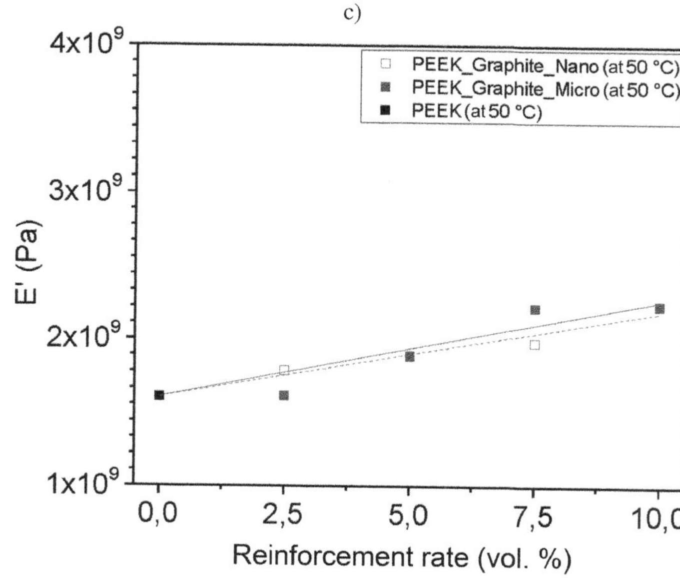

d)

Figure 4.16. *Evolution of the conservation modulus E' at 50° C for each reinforcement: a) silicon carbide, b) alumina, c) boron nitride and d) graphite. For a color version of this figure, see www.iste.co.uk/bai/nanocomposites.zip*

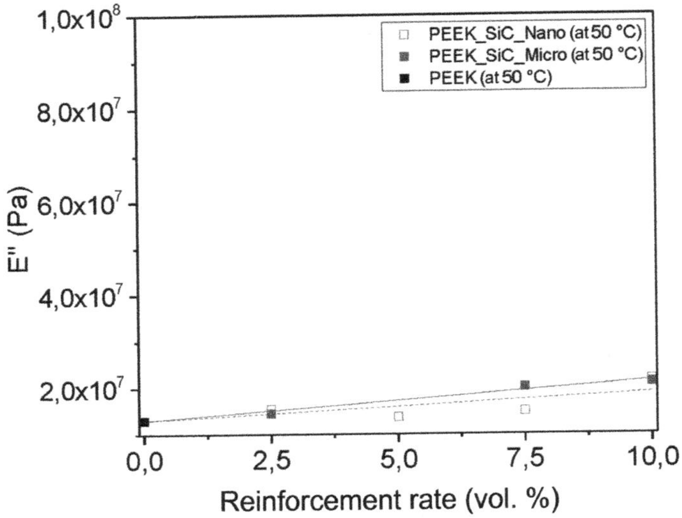

1,0x10^8
8,0x10^7
6,0x10^7
4,0x10^7
2,0x10^7
E'' (Pa)
PEEK_SiC_Nano (at 50 °C)
PEEK_SiC_Micro (at 50 °C)
PEEK (at 50 °C)
0,0
2,5
5,0
7,5
10,0
Reinforcement rate (vol. %)

a)

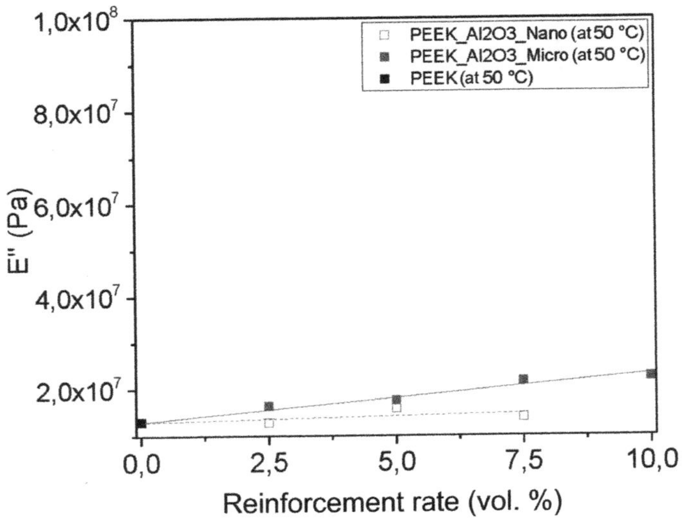

1,0x10^8
8,0x10^7
6,0x10^7
4,0x10^7
2,0x10^7
E'' (Pa)
PEEK_Al2O3_Nano (at 50 °C)
PEEK_Al2O3_Micro (at 50 °C)
PEEK (at 50 °C)
0,0
2,5
5,0
7,5
10,0
Reinforcement rate (vol. %)

b)

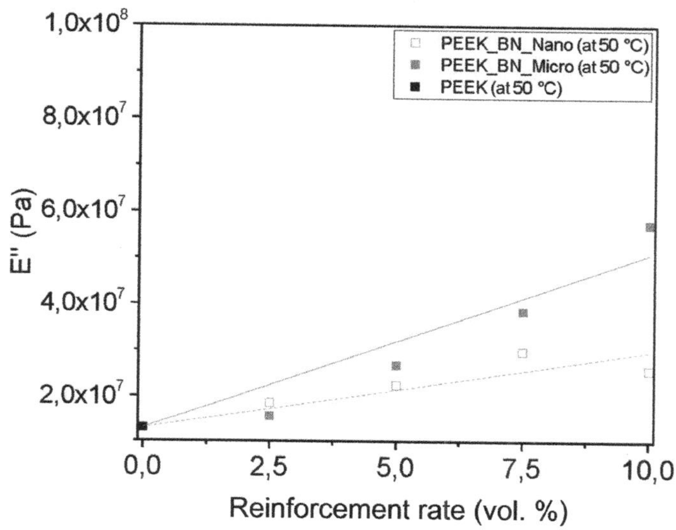

c)

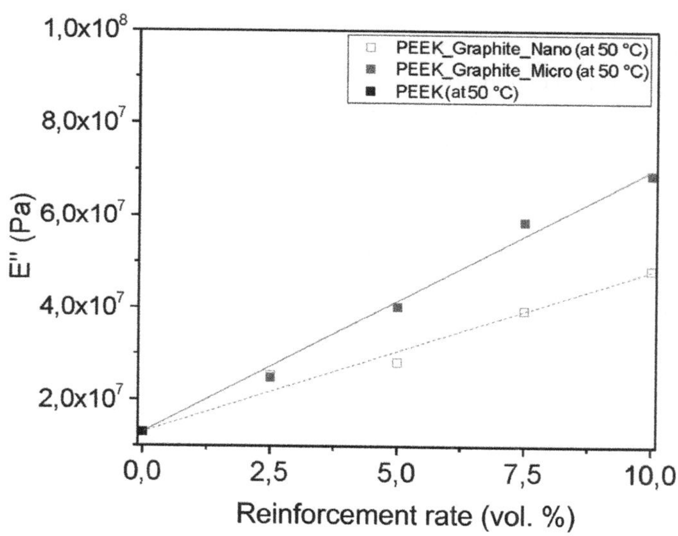

d)

Figure 4.17. *Evolution of loss modulus E'' at 50 °C for each reinforcement: a) silicon carbide, b) alumina, c) boron nitride and d) graphite. For a color version of this figure, see www.iste.co.uk/bai/nanocomposites.zip*

Here again, BN stands out with a possibly prominent interphase role. In this interphase, the molecules are organized differently from the molecules in the volume, due to the interaction between the polymer chains and the surface of the reinforcement. The surface of the BN sheets is devoid of chemical groups capable of interacting with the polymer chain, which is why this material is called chemically inert. The BN/polymer interface is therefore not adherent, which leaves more mobility for the macromolecules, resulting in a decrease in mechanical stiffness at low levels. At higher rates, this tendency is compensated by the stiffness of the fillers themselves because the Young modulus of the fillers is much higher than that of the matrix. The mixing rule then becomes applicable.

The presence of this interphase in the vicinity of the fillers leads to a gain or loss of mobility for the macromolecules in the interphase. This change in mobility is measurable by rheometry, and can be seen in the $tan\ \delta$ curves shown in Figures 4.18(a) and (b). For most composites, $tan\ \delta$ is less than or equal to PEEK at the glass transition temperature, except for 10 volume % SiC nanofiller-reinforced samples and 5 volume % Al_2O_3 microfiller. For these two samples, the loss modulus E'' is larger than the storage modulus E'. The energy dissipated due to molecular friction is higher.

The curvature that occurs at the beginning of the glass transition of the $tan\ \delta$ curves corresponds to the mobility of the macromolecules and can therefore be related to interfacial interactions. Since the shape of the curves is very similar for all fillers and proportions, let us focus on the curvature of the foot of the peak: for all SiC and alumina ratios, the glass transitions occur at the same temperature. Figures 4.19(c) and (d) show that the molecular mobility of the polymer reinforced with micrometric alumina is identical, regardless of the reinforcement rate.

On the contrary, for lamellar fillers, the glass transition starts at a lower temperature as the rate increases. At the interphase level, the bonds between the reinforcements and the matrix are weaker, as previously explained; we can conclude that adhesion between the reinforcement and the matrix is less intense. In addition, in the case of poor interfacial adhesion, stresses become concentrated at the interface and microcracks appear at the interface and propagate, leading to material failure. A good interfacial adhesion allows the transmission of mechanical stresses from the matrix to the filler; there is no interruption, the material will be more resistant to failure.

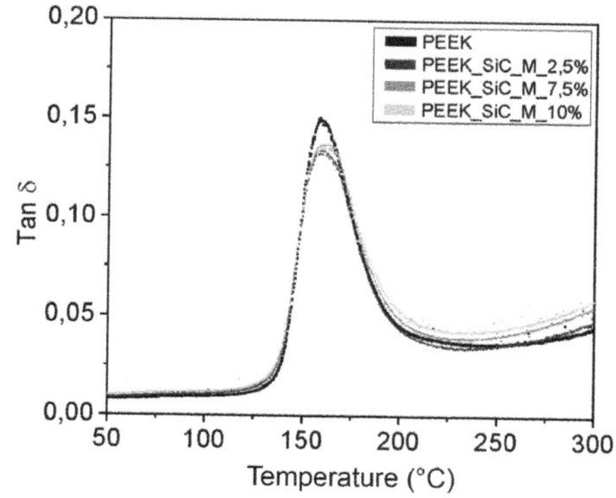
PEEK
PEEK_SiC_M_2,5%
PEEK_SiC_M_7,5%
PEEK_SiC_M_10%
Tan δ
Temperature (°C)

a)

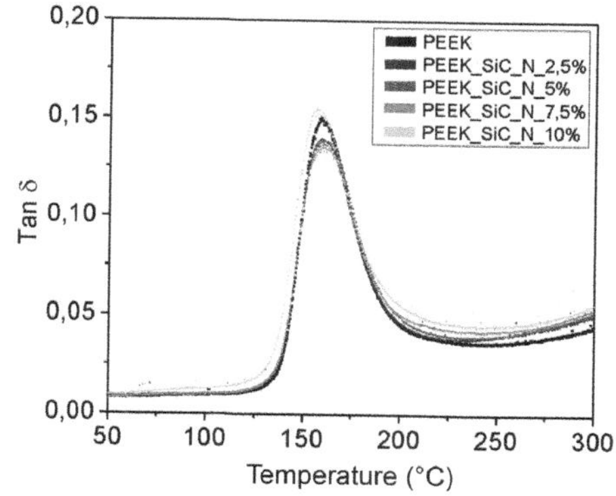
PEEK
PEEK_SiC_N_2,5%
PEEK_SiC_N_5%
PEEK_SiC_N_7,5%
PEEK_SiC_N_10%
Tan δ
Temperature (°C)

b)

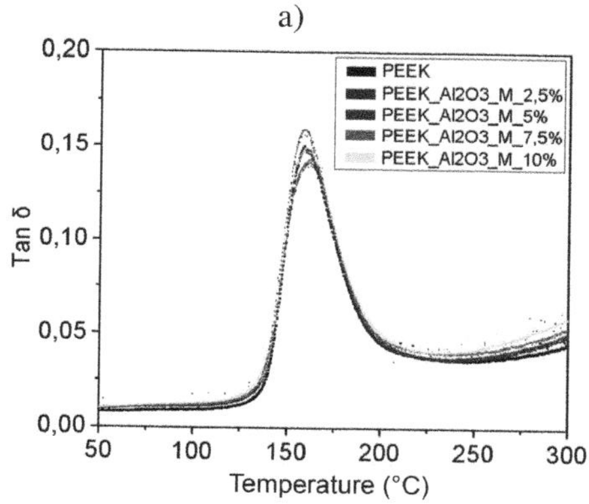
PEEK
PEEK_Al2O3_M_2,5%
PEEK_Al2O3_M_5%
PEEK_Al2O3_M_7,5%
PEEK_Al2O3_M_10%
Tan δ
Temperature (°C)

c)

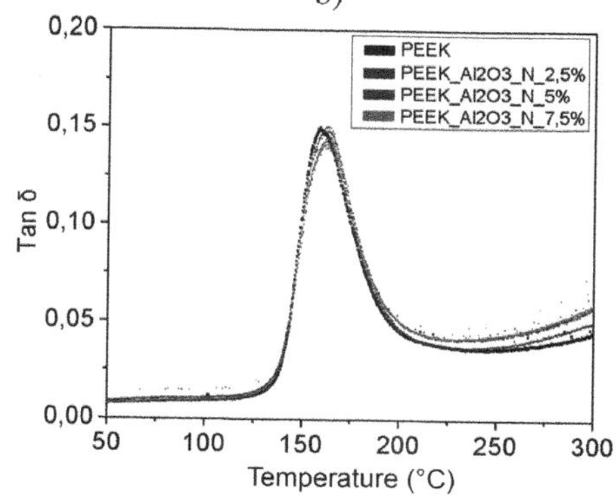
PEEK
PEEK_Al2O3_N_2,5%
PEEK_Al2O3_N_5%
PEEK_Al2O3_N_7,5%
Tan δ
Temperature (°C)

d)

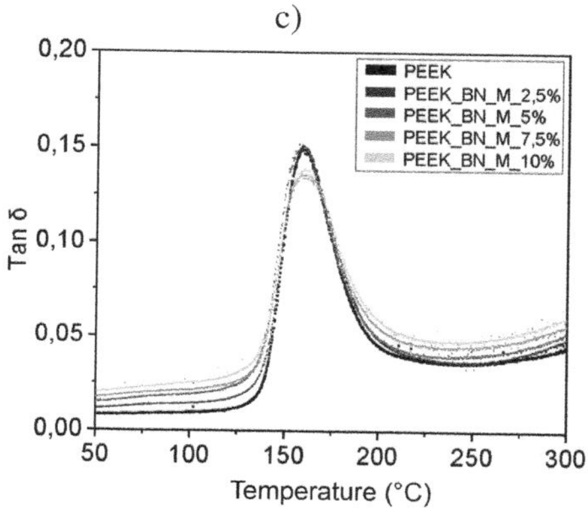
PEEK
PEEK_BN_M_2,5%
PEEK_BN_M_5%
PEEK_BN_M_7,5%
PEEK_BN_M_10%
Tan δ
Temperature (°C)

e)

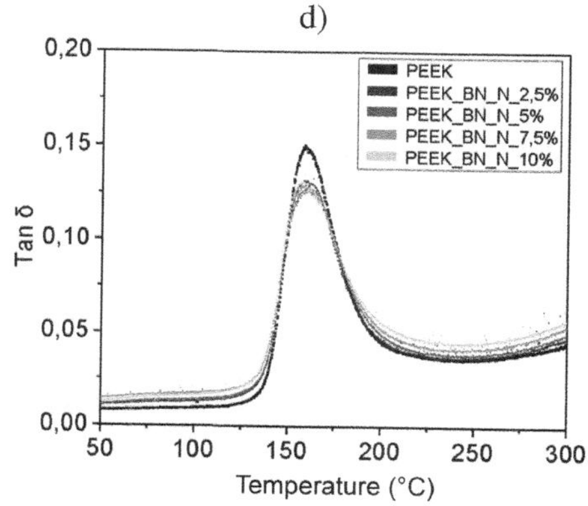
PEEK
PEEK_BN_N_2,5%
PEEK_BN_N_5%
PEEK_BN_N_7,5%
PEEK_BN_N_10%
Tan δ
Temperature (°C)

f)

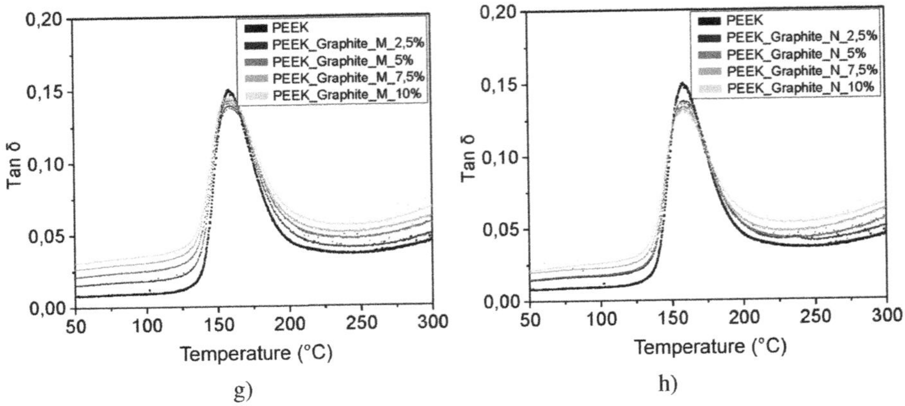

Figure 4.18. *Evolution of tan δ for PEEK and a) and b) silicon carbide, c) and d) alumina and e) and f) boron nitride reinforced samples for micrometer-sized (a, c, e) and nanometer-sized (b, d, f) fillers, and graphite reinforced samples for g) micrometer and h) nanometer sized fillers. The deformation is 0.5% at 1 Hz. For a color version of this figure, see www.iste.co.uk/bai/nanocomposites.zip*

4.3.6.2. *Tribological study*

A tribological test campaign is conducted using a commercial tribometer in a plane/ball configuration. The ball follows a reciprocal translation motion over a length of 10 mm and with a frequency of 5 Hz. A force of 25 N is applied. The friction coefficients are measured from the ratio of the tangential force to the applied force. The traces are measured and then observed to identify accommodation modes and the presence of debris.

In Figure 4.20, three groups of composites can be highlighted by comparing the depth of the traces and their friction coefficients. The group of materials at the bottom left represents *moderate surface wear* along with reduced energy dissipation. It brings together PEEK and composites based on 2.5 and 5 volume % graphite microparticles, as well as one containing 2.5 volume % graphite nanoparticles. Two other groups correspond to *high* and to *severe wear*. This last group contains traces with depths greater than 125 μm. Composites with h-BN and contents of 2.5 volume % of microparticles or 2.5 volume % and 5 volume % of nanoparticles are among the materials showing this severe wear. PEEK filled with 7.5 and 10 volume % of graphite, regardless of the filler size, also fall within this group.

When materials have a coefficient of friction between 0.3 and 0.45, then they dissipate little energy during tribological testing, and this leads to a small wear trace. When $\mu > 0.45$, the traces are large in size.

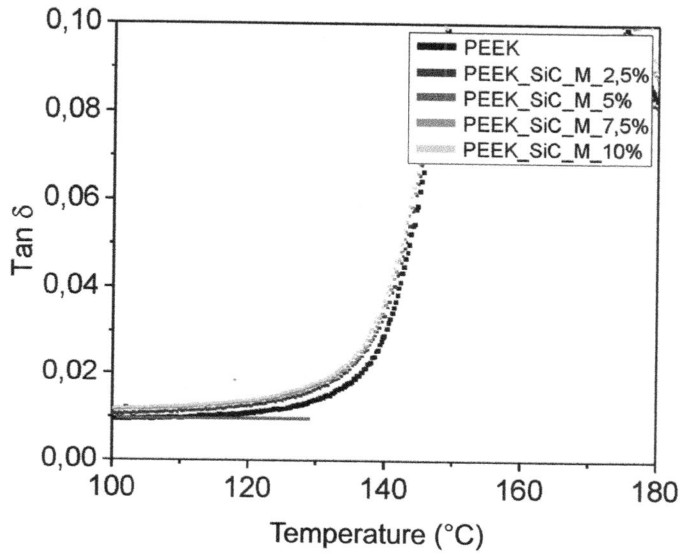
PEEK
PEEK_SiC_M_2,5%
PEEK_SiC_M_5%
PEEK_SiC_M_7,5%
PEEK_SiC_M_10%
Tan δ
Temperature (°C)

a)

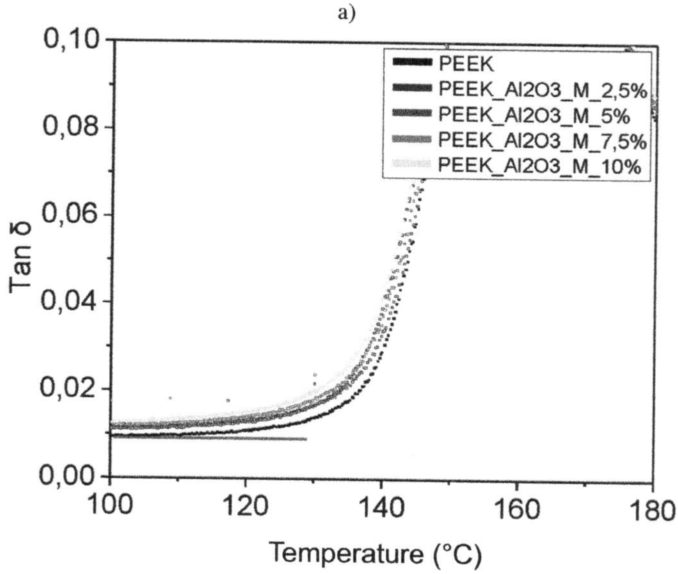
PEEK
PEEK_Al2O3_M_2,5%
PEEK_Al2O3_M_5%
PEEK_Al2O3_M_7,5%
PEEK_Al2O3_M_10%
Tan δ
Temperature (°C)

b)

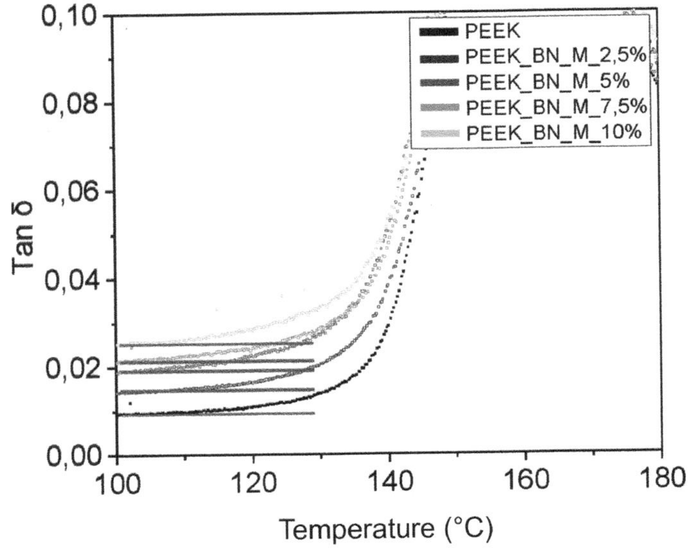

c)

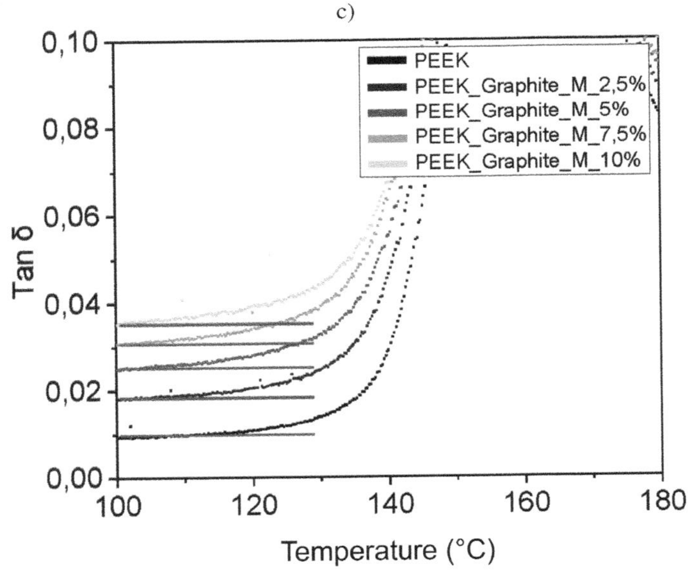

d)

Figure 4.19. *Evolution of tan δ for PEEK and reinforced samples: with a) silicon carbide, b) alumina, c) boron nitride and d) with micrometric size graphite. The deformation is 0.5% at 1 Hz. For a color version of this figure, see www.iste.co.uk/bai/nanocomposites.zip*

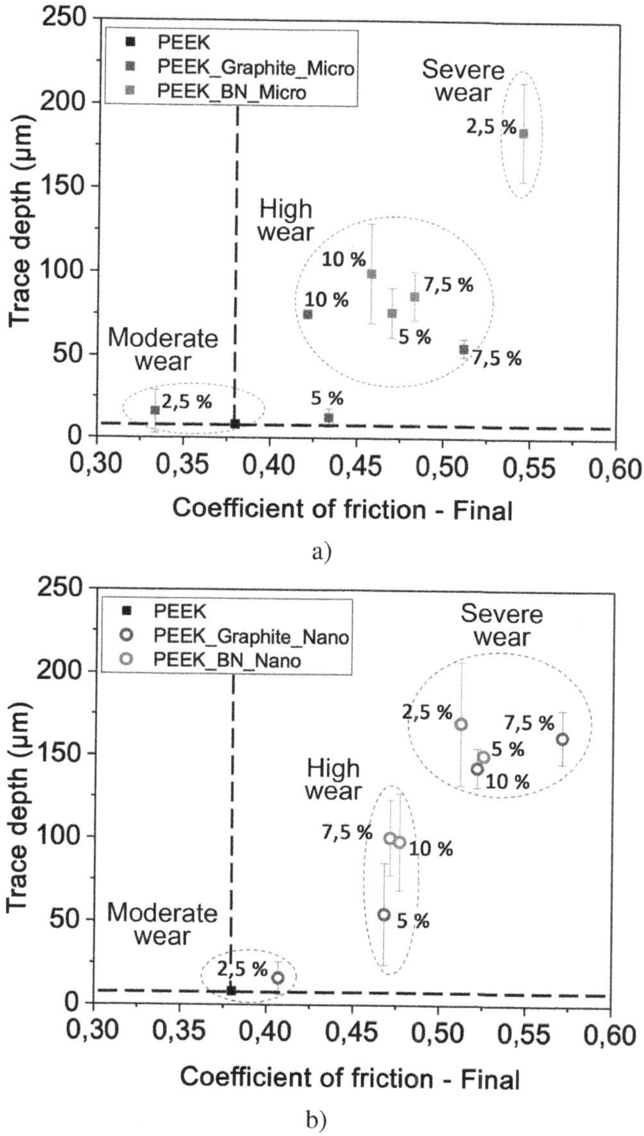

Figure 4.20. *Evolution of depth according to the coefficient of friction for PEEK and lamellar particle-filled composites of a) micrometer and b) nanometer sizes. For a color version of this figure, see www.iste.co.uk/bai/nanocomposites.zip*

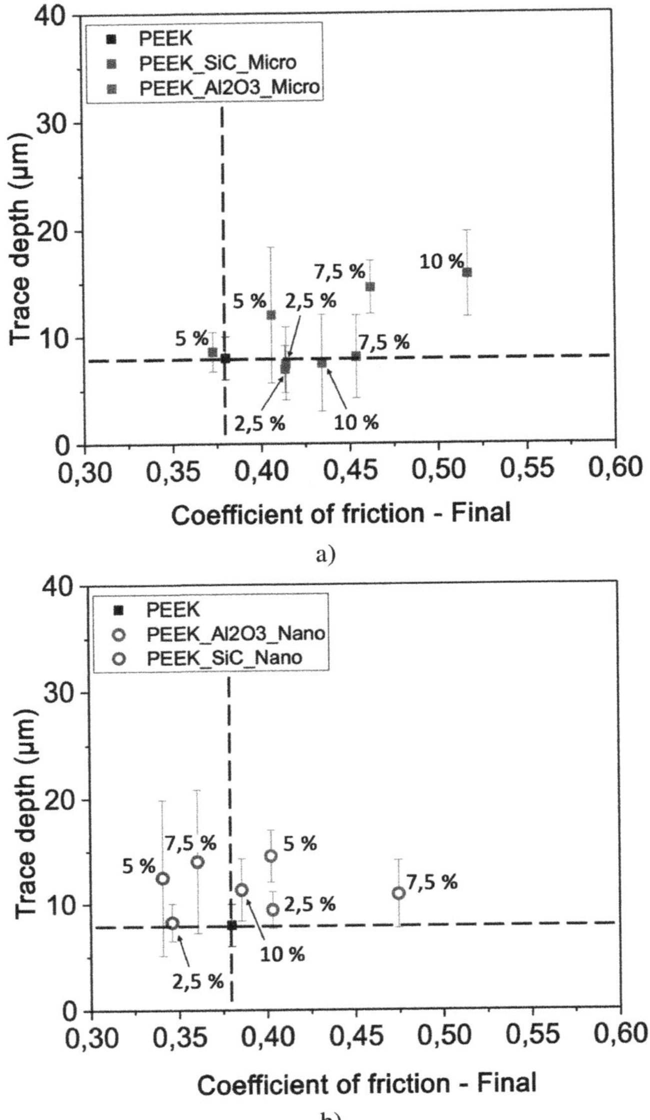

Figure 4.21. *Evolution of depth according to the coefficient of friction for PEEK and composites filled with particulate fillers of a) micrometer and b) nanometer sizes. For a color version of this figure, see www.iste.co.uk/bai/nanocomposites.zip*

Composites with the lowest graphite content have the lowest coefficient of friction and the smallest trace sizes. The addition of h-BN leads to the highest friction coefficients and the most severe wear.

With PEEK added with SiC or Al_2O_3 particles, the trace depths and friction coefficients are reduced and clustered around the PEEK values (Figure 4.21). This corresponds to the area that we have associated with *moderate wear*.

The materials have therefore retained their mechanical strength with the addition of alumina and SiC fillers, limiting the opening and deepening of the groove caused by the movement of the ball. Nevertheless, we verified the condition of the steel balls after the tribological tests and identified the presence of brown debris and scratches on their surfaces. The balls are therefore slightly damaged and emit material that oxidizes within the contact interface. Alumina and SiC particles, therefore, contribute to the degradation of the steel balls with which the composites are made to slide.

The modes of accommodation that prevail in these steel/PEEK matrix composites friction pairs are essentially plastic deformation modes. Depending on the type of filler added, they can be intense and produce high or severe wear. This is the case with BN. The accommodation modes are also dependent on the lamellar particle content and, in general, the higher the content, the more pronounced the plastic deformation.

With energy dissipation comparable to PEEK, the addition of Al_2O_3 or of SiC significantly reduces wear. In this case, the composite does not collapse under tribological loading, which is in good agreement with a reduced variation of the storage modulus E', loss modulus E'' and damping factor $\tan\delta$ for composites. Indeed, the complex moduli of the composites are very close to the moduli of the PEEK matrix. The plastic deformation in these composites is limited, but this reinforcement has a counterpart: the steel balls are slightly degraded.

Ultimately, optimized configurations can be proposed by employing micrometric graphite-filled PEEK composites at a volume ratio of less than 5%. This formulation would have the advantage of avoiding the emission of debris by the ball. Composites prepared with SiC or alumina will remarkably reduce wear compared to composites containing BN. It should be noted that particle size will not produce a significant effect on wear for these types of fillers. On the other hand, the use of SiC nanoparticles, between 2.5 and 7.5 volume %, will induce a decrease in the friction coefficient below that of PEEK and thus a lower dissipated energy during sliding.

4.4. Conclusion

Understanding the effect of size and filler/matrix interactions in nanocomposites is crucial to develop high-performance and durable materials. Keeping in mind that the role of the interphase on the physical properties of materials is undeniable, it remains difficult to predict these properties that differ for each couple filler/matrix. We sought to bring forward the influence of the size and nature of the fillers in microcomposite and PEEK matrix-based nanocomposite materials. For this purpose, samples loaded with SiC, alumina, graphite and BN were prepared using the melt process, using volume rates between 2.5 and 10%. The fillers are commercial and can be found in powder form, with each type of filler being classified as nanometric and micrometric in size. The initial particle size distribution was verified by laser granulometry: the distributions are wide and agglomerates are initially present, and all the more for nanoparticles. At the end of the melt mixing operation, the distribution of particles in the matrix was verified using SEM and image analysis. Here again, agglomerates were observed, but their size is smaller, especially for nanometric particles, which shows the efficiency of shear during mixing to break the aggregates and disperse the particles. As desired, the filler/matrix interface is more important in the case of nanocomposites.

The prepared composites were characterized by differential scanning calorimetry, thermal measurements, dynamic mechanical analysis and finally by frictional resistance measurements. The glass transition of PEEK, which reflects the mobility of the macromolecules, decreases by a few degrees in the presence of SiC and alumina for the highest levels, suggesting a stiffening of polymeric chains. Nano-sized fillers do not affect the matrix mobility more than micrometer sizes, despite the larger contact area. For graphite and BN, the results are more scattered, with deviations of 10°C with respect to the reference PEEK, which does not allow us to conclude about the existence of an interphase. Despite the glass transition being the expression of cooperative motions, it is likely that the mobility of macromolecules is inhomogeneous in the material, with localized less rigid areas. The crystallinity, initially 35% for PEEK, decreases for higher concentrations regardless of filler type. This decrease is more important for SiC, which reflects a hindrance caused by the surface of the particles, for the macromolecules to self-organize. The creation of weak bonds between the SiC and polymer chain could be at the origin of this problem.

For all fillers, the thermal conductivity of PEEK is slightly increased, due to the high conductivity of the selected particles. However, the increase does not follow a linear relationship with the concentration, which confirms the complexity of the heat transfer phenomenon that results from the movement of electrons and phonons. For graphite, this increase is noticeable and seems to mark the appearance of a percolating network, even if the concentrations studied remain too low to confirm it. The micrometric particles seem to bring a slight gain in heat transfer compared to the nanometric particles. The presence of large particles that constitute high conductivity zones that act as thermal bridges in the material could explain these results.

The dynamic mechanical analysis in the linear viscoelastic regime highlights the reinforcement provided by the particles, which themselves have much larger Young's moduli than the PEEK matrix. The increase in the storage modulus E' is almost linear with the volume rate, except for BN and graphite. Indeed, the evolution of their modulus E' follows a linearity without passing through the PEEK value. BN is inert and does not establish chemical bonds with macromolecules. It is likely that adhesion is non-existent at the interface, suggesting an interphase effect in which the molecules would have more mobility. This gain in mobility was not shown by DSC, but because of the ratio E'/E'', that is to say, of the tangent loss factor δ. The curves representing the loss factor are very similar for all materials and similar in shape to that of PEEK, which shows the absence of a population of chains with drastically different behavior. Nevertheless, a close study of the loss factor peak shows that the curvature marking the onset of the glass transition appears a few degrees earlier for BN and graphite, which supports this hypothesis. Surprisingly, this interphase effect seems to be more pronounced for micrometric particles than for nanometric ones.

Finally, the effect of the size and nature of the fillers on the frictional properties was studied. A size effect is undeniable for SiC and alumina: their friction coefficient is lower for nanometric fillers. Therefore, the energy dissipated during friction in these materials will be less, which allows for energy savings during use. Concerning their wear, measured here by the shape of the wear trace, the nature of the load predominates. A distinction must be made between lamellar fillers – graphite and BN – which cause high wear, and particulate fillers – SiC and alumina – for which wear is low.

Finally, this work shows that nanometric sizes do not bring the expected gain on thermal and mechanical properties, but allow a better understanding of the role of filler/matrix interactions. The effect of the nature of the filler and in particular of the interfacial interaction, and namely of the bonds established between the particles and the macromolecules, seems to be dominant over the size effect. Improving the dispersion and exfoliation state of the fillers is essential to increase the contact surface and thus benefit more from the effects of the filler/matrix interphase. However, by using the melt mixing process and the materials studied, it will be difficult to do better due to the high viscosity of PEEK and its sensitivity to thermo-oxidative degradation.

4.5. References

Attwood, T.E., Dawson, P.C., Freeman, J.L., Hoy, L.R.J., Rose, J.B., Staniland, P.A. (1981). Synthesis and properties of polyaryletherketones. *Polymer*, 22(8), 1096–1103 [Online]. Available at: https://linkinghub.elseviercom/retrieve/pii/0032386181902998.

Bedi, H.S., Tiwari, M., Agnihotri, P.K. (2018). Quantitative determination of size and properties of interphases in carbon nanotube-based multiscale composites. *Carbon*, 132, 181–190 [Online]. Available at: https://linkinghub.elsevier.com/retrieve/pii/ S0008622318301830.

Bernal, J. (1924). The structure of graphite. *Proceedings of the Royal Society A*, 106, 749.

Chen, E.J.H. and Hsiao, B.S. (1992). The effects of transcrystalline interphase in advanced polymer composites. *Polymer Engineering & Science*, 32(4), 280–286 [Online]. Available at: https://onlinelibrary.wiley.com/doi/abs/10.1002/pen.760320408.

Choi, J., Shin, H., Yang, S., Cho, M. (2015). The influence of nanoparticle size on the mechanical properties of polymer nanocomposites and the associated interphase region: A multiscale approach. *Composite Structures*, 119, 365–376 [Online]. Available at: https://linkinghub.elsevier.com/retrieve/pii/S0263822314004474.

Crosby, A.J. and Lee, Y. (2007). Polymer nanocomposites: The "nano" effect on mechanical properties. *Polymer Reviews*, 47, 217–229.

David, L. and Etienne, S. (1992). Molecular mobility in para-substituted polyaryls 1. Sub-(Tg) relaxation phenomena in poly(aryl ether ether ketone). *Macromolecules*, 25(17), 4302–4308 [Online]. Available at: https://pubs.acs.org/doi/abs/10.1021/ma00043a010.

Decup, M. (2010). Impact des procédés industriels de traitement sur les propriétés diélectriques des substrats d'alumine utilisés dans les modules de commutation haute tension. PhD Thesis, Université Toulouse 3 Paul Sabatier, Toulouse.

Díez-Pascual, A.M., Guan, J., Simard, B., Gómez-Fatou, M.A. (2012). Poly(phenylene sulphide) and poly(ether ether ketone) composites reinforced with single-walled carbon nanotube buckypaper II. Mechanical properties, electrical and thermal conductivity. *Composites Part A: Applied Science and Manufacturing*, 43(6), 1007–1015 [Online]. Available at: https://linkinghub.elsevier.com/retrieve/pii/S1359835X11003629.

Domenech, T., Zouari, R., Vergnes, B., Peuvrel-Disdier, E. (2014). Formation of fractal-like structure in organoclay-based polypropylene nanocomposites. *Macromolecules*, 47(10), 3417–3427 [Online]. Available at: https://pubs.acs.org/doi/10.1021/ma5001354.

Doumeng, M., Makhlouf, L., Berthet, F., Marsan, O., Delbé, K., Denape, J., Chabert, F. (2021). A comparative study of the crystallinity of polyetheretherketone by using density, DSC, XRD, and Raman spectroscopy techniques. *Polymer Testing*, 93, 106878 [Online]. Available at: https://linkinghub.elsevier.com/retrieve/pii/ S0142941820321073.

Drzal, L.T., Rich, M.J., Koenig, M.F., Lloyd, P.F. (1983). Adhesion of graphite fibers to epoxy matrices II. The effect of fiber finish. *The Journal of Adhesion*, 16, 133–152.

Falin, A., Cai, Q., Santos, E., Scullion, D., Qian, D., Zhang, R., Yang, Z., Huang, S., Watanabe, K., Taniguchi, T., Barnett, M., Chen, Y., Ruoff, R., Li, L. (2017). Mechanical properties of atomically thin boron nitride and the role of interlayer interactions. *Nature Communications*, 8(1) [Online]. Available at: http://www.nature.com/articles/ ncomms15815 and https://creativecommons.org/licenses/by/4.0/deed.fr.

Flory, P.J. (1953). *Principles of Polymer Chemistry*. Cornell University Press, New York.

Goyal, R., Tiwari, A., Mulik, U., Negi, Y. (2007). Novel high performance Al_2O_3/poly(ether ether ketone) nanocomposites for electronics applications. *Composites Science and Technology*, 67(9), 1802–1812 [Online]. Available at: https://linkinghub.elsevier.com/retrieve/pii/S0266353806004143.

Jesson, D.A. and Watts, J.F. (2012). The interface and interphase in polymer matrix composites: Effect on mechanical properties and methods for identification. *Polymer Reviews*, 52, 321–354.

Kadiyala, A.K. and Bijwe, J. (2018). Poly (ether ether ketone)-silicon carbide composite adhesives for elevated temperature applications of stainless steel joints. *Composites Science and Technology*, 155, 177–188 [Online]. Available at: https://linkinghub.elsevier.com/retrieve/pii/S0266353817324442.

Kadiyala, A.K., Bijwe, J., Kalappa, P. (2018). Investigations on influence of nano and micron sized particles of SiC on performance properties of PEEK coatings. *Surface and Coatings Technology*, 334, 124–133 [Online]. Available at: https://www.sciencedirect.com/science/article/pii/S0257897217311726.

Karger-Kocsis, J., Mahmood, H., Pegoretti, A. (2015). Recent advances in fiber/matrix interphase engineering for polymer composites. *Progress in Materials Science*, 73, 1–43 [Online]. Available at: https://linkinghub.elsevier.com/retrieve/pii/S0079642515000328.

Khostavan, S., Fazli, M., Ahangari, M.G., Rostamiyan, Y. (2019). The effect of interaction between nanofillers and epoxy on mechanical and thermal properties of nanocomposites: Theoretical prediction and experimental analysis. *Advances in Polymer Technology*, 8156718.

Kronberg, M.L. (1957). Plastic deformation of single crystals of sapphire: Basal slip and twinning. *Acta Metallurgica*, 5(507), 507–524.

Kuo, M.C., Tsai, C.M., Huang, J.C., Chen, M. (2005). PEEK composites reinforced by nano-sized SiO_2 and Al_2O_3 particulates. *Materials Chemistry and Physics*, 90(1), 185–195 [Online]. Available at: http://www.sciencedirect.com/science/article/pii/S0254058404005218.

Lafon-Placette, S. (2015). Performances tribologiques d'un carbure de silicium pour paliers d'étanchéité dynamique fonctionnant en conditions sévères. PhD Thesis, Institut National Polytechnique, Toulouse.

Lambrecht, W.R.L., Limpijumnong, S., Rashkeev, S., Segall, B. (1997). Electronic band structure of SiC polytypes: A discussion of theory and experiment. *Phys. Stat. Sol. B*, 202(1), 5–33 [Online]. Available at: https://onlinelibrary.wiley.com/doi/abs/10.1002/1521-3951%28199707%29202%3A1%3C5%3A%3AAID-PSSB5%3E3.0.CO%3B2-L.

Lee, M.-Y., Wang, T.-Y., Tsai, J.-L. (2016). Mechanical properties of nanocomposites with functionalized graphene. *Journal of Composite Materials*, 50(27), 3779–3789.

Lefevre, J. (2008). Étude des effets d'irradiation dans le polytype cubique du carbure de silicium par les techniques spectroscopiques de photoluminescence et de résonance paramagnétique électronique. PhD Thesis, École Polytechnique, Palaiseau.

Liu, Y. and Kumar, S. (2014). Polymer/carbon nanotube nano composite fibers: A review. *ACS Applied Materials & Interfaces*, 6(9), 6069–6087 [Online]. Available at: https://pubs.acs.org/doi/10.1021/am405136s.

Normand, G., Peuvrel-Disdier, E., Vergnes, B. (2016). Matrix degradation during high speed extrusion of polypropylene/clay nanocomposites – Influence on filler dispersion. *International Polymer Processing*, 31(4), 508–516 [Online]. Available at: https://www.degruyter.com/view/journals/ipp/31/4/article-p508.xml.

Panda, J.N., Bijwe, J., Pandey, R.K. (2019). Optimization of graphite contents in PAEK composites for best combination of performance properties. *Composites Part B: Engineering*, 174 [Online]. Available at: https://linkinghub.elsevier.com/retrieve/pii/S1359836819311709.

Pauling, L. (1932). The nature of the chemical bond IV, the energy of single bonds and the relative electronegativity of atoms. *Journal of American Chemical Society*, 54(9), 3570–3582.

Rae, P.J., Brown, E.N., Orler, E.B. (2007). The mechanical properties of poly(ether-ether-ketone) (PEEK) with emphasis on the large compressive strain response. *Polymer*, 48(2), 598–615 [Online]. Available at: https://linkinghub.elsevier.com/retrieve/pii/S0032386106012742.

Ramsdell, L. (1947). Studies on silicon carbide. *American Mineralogist*, 32, 19.

Roussel, N. (2013). Optimisation du dopage d'alumines nanométriques et frittage par SPS. Application aux céramiques transparentes. PhD Thesis, Université Toulouse 3 Paul Sabatier, Toulouse.

Saleem, H., Edathil, A., Ncube, T., Pokhrel, J., Khoori, S., Abraham, A., Mittal, V. (2016). Mechanical and thermal properties of thermoset-graphene nanocomposites. *Macromolecular Materials and Engineering*, 301(3), 231–259 [Online]. Available at: http://doi.wiley.com/10.1002/mame.201500335.

Savage, R. and Schaefer, D. (1956). Vapor lubrication of graphite sliding contacts. *Journal of Applied Physics*, 27(2), 136–138.

Shiu, S.-C. and Tsai, J.-L. (2014). Characterizing thermal and mechanical properties of graphene/epoxy nanocomposites. *Composites Part B: Engineering*, 56, 691–697 [Online]. Available at: https://linkinghub.elsevier.com/retrieve/ pii/S1359836813005209.

Talbi, F. (2010). Étude des phénomènes de conduction et de rupture diélectrique dans les céramiques à haute teneur en alumine. PhD Thesis, Université Mouloud Mammeri, Tizi-Ouzou.

Termoss, H. (2009). Préparation de revêtements de nitrure de bore (BN) par voie polymère précéramique : étude des paramètres d'élaboration : caractérisations physico-chimiques. PhD Thesis, Université Claude Bernard–Lyon 1, Lyon.

Terrones, M., Martín, O., González, M., Pozuelo, J., Serrano, B., Cabanelas, J.C., Vega-Díaz, S.M., Baselga, J. (2011). Interphases in graphene polymer-based nanocomposites: Achievements and challenges. *Advanced Materials*, 23(44), 5302–5310 [Online]. Available at: http://doi.wiley.com/10.1002/adma.201102036.

Yacoby, A. (2011). Graphene: Tri and tri again. *Nature Physics*, 7(12), 925–926 [Online]. Available at: http://www.nature.com/articles/nphys2166.

Zaidi, H., Robert, F., Paulmier, D. (1995). Influence of adsorbed gases on the surface energy of graphite: Consequences on the friction behaviour. *Thin Solid Films*, 264(1), 46–51 [Online]. Available at: https://linkinghub.elsevier.com/retrieve/pii/0040609095066063.

Zouari, R., Domenech, T., Vergnes, B., Peuvrel-Disdier, E. (2012). Time evolution of the structure of organoclay/polypropylene nanocomposites and application of the time-temperature superposition principle. *Journal of Rheology*, 56(4), 725–742 [Online]. Available at: http://sor.scitation.org/doi/10.1122/1.4708602.

5

Study of Interface and Interphase between Epoxy Matrix and Carbon-based Nanofillers in Nanocomposites

Yu LIU, Delong HE, Ann-Lenaig HAMON and Jinbo BAI
CentraleSupélec, Paris-Saclay University, Gif-sur-Yvette, France

Our study focuses on the characterization of the chemical nature and mechanical properties of the interface created by carbon micro- and nanofillers in an epoxide matrix. A direct measurement of the materials structure has been carried out on a nanometric scale using electron microscopy. Electron energy-loss spectroscopy (EELS) performed in a scanning transmission electron microscope (STEM) measured the plasmon peak position in the low-loss region, which is related to the local materials density. *In situ* tensile tests were carried out in a scanning electron microscope (SEM) on samples machined using a focused ion beam (FIB). These experiments were conducted on several types of fillers: oxidized or non-oxidized graphite platelets, carbon fibers (CFs) with/without their sizing layers and with/without nanotubes grown on their surfaces.

5.1. Introduction

Composite materials are known for combining a low weight with interesting properties such as mechanical strength. In this chapter, we report our investigations on composites where both the matrix and the fillers were carbon based. The matrix was a polymer (epoxide), chosen for its ease in use, whereas the different fillers we

Nanocomposites,
coordinated by Jinbo BAI.

studied were also carbonaceous and either microsized (CFs) or nanosized (nanotubes, graphite nanoplatelets).

The understanding of the structure at the interface between the filler and the matrix is essential to control the properties of the homogenized, macroscopic material. Thereby near the interface, the matrix is reported to experience a modification of its structure, the interphase (Pukànszky 2005). This structural alteration can also locally modify the material properties and influence the composite behavior on a macroscopic scale such as, for instance, a decreased composite strength. Delale and Erdogan (1988) proposed a model taking into account a continuous variation in the interphase of the elastic moduli.

Besides, we seek to improve the material properties at the interface, including a modification of the filler surface using various strategies that are described below, such as filler oxidation or the synthesis of hybrid materials.

We carried out a direct measurement of the material local structure at the interface, accounting for both the presence of an interphase – whose thickness must be determined – and the surface modification. These observations were coupled to an investigation of the material strength, here again in the interface vicinity. The fact that both the matrix and the sample were carbonaceous gave rise to difficulties in such investigations, since we must use appropriate characterization techniques that can discriminate the different phases in the composite on a nanometer scale (Wu *et al.* 2014).

5.1.1. *Surface modification of the fillers*

Two strategies prevail in order to improve the cohesion at the interface, which both imply a modification of the filler/matrix surface. The first one is physical and is related to an increase in the surface roughness (Im and Kim 2012). The second one consists of oxidizing the filler in order to bridge the matrix (Sengupta *et al.* 2011).

The samples studied here were made alternately, using one of these methods. So, with the first method, carbon nanotubes were grown on CFs in order to increase the interacting surface (Saba *et al.* 2013). Otherwise, graphite nanoplatelets were oxidized – as previously carried out by Ramanathan *et al.* (2008), who distinguished strong interfaces from weak interfaces – or CFs were studied without the grown nanotubes mentioned above, but with their sizing layers as commercially provided, or in another case, the carbon nanotubes grown on the CFs were then oxidized. These composites obtained with modified fillers were also compared with composites with non-treated fillers, that is, non-oxidized graphite nanoplatelets and CFs after the removal of their sizing layer.

5.1.2. *Experimental techniques*

Given the filler treatment, it was necessary to investigate the composite structure at the scale of the filler-matrix interface, in order to access the structure modification generated by the treatment in comparison with the untreated case. Transmission electron microscopy (TEM), used in a scanning mode (STEM), provides a multitude of information which, moreover, has the advantage of being multimodal, for instance, by coupling high-resolution imaging and EELS. This characterization method gives information not only on the chemical nature of the elements, but also on their bonding states or the local valence electron density.

Besides, since the mechanical properties of the composite are at stake, tensile mechanical tests were carried out on a microsample centered on an interface between different types of a modified CF and the epoxide matrix. In order to do so, the sample was first milled with a FIB in an SEM after selecting the proper area. The tensile load was applied using the microscope nanomanipulator.

5.2. Structural analysis of the interface with electron energy-loss spectroscopy

5.2.1. *Analysis technique*

The TEM analyses were carried out using an FEI Titan G2 80 – 300 microscope with a high brightness field emission gun (X-FEG). In order to protect the sample from the irradiation effects, the observations and analyses were carried out with the lowest possible high tension, namely 80 kV. As a matter of fact, spectra acquired with different high tensions showed an alteration of the observed structure after 1 minute of irradiation at 200 kV, whereas the features in the spectra acquired at 80 kV remained stable after the same amount of time. At this high tension, the spatial resolution is then around 0.125 nm, and for the EELS analyses, the energy resolution is around 0.8 eV.

EEL spectra contain various information depending on the type of inelastic interactions between the probing electrons and the sample (Egerton 1986). In this study, we focused on two types of information present in two different energy-loss ranges. In the core-loss region (100 – 1500 eV), the ionization edge gives the chemical composition in the probed area. The energy-loss near-edge spectrum (ELNES) study of the carbon K-edge at 284 eV was more specifically studied: the energy resolution allows access to the fine structure, and thus the chemical bonding of the carbon atoms in the probed area. In the low-loss range (0 – 100 eV), the plasmon peak is related to the local valence electron density, as explained in Ferrari *et al.* (2000) and summed-up below.

The plasmon peak position in the spectra E_p is related to the local density of valence electrons n_v:

$$E_p = \hbar e \sqrt{\frac{n_v}{\epsilon_0 m^*}}$$

where $\hbar$ stands for the reduced Planck's constant, e is the elementary charge, ϵ_0 is the vacuum permittivity and m^* is the electron effective mass.

As a carbon atom owns four valence electrons, one can thus infer the material density n in a pure carbon material, as a number of atoms per cubic meter:

$$n = \frac{\epsilon_0 m^*}{4\hbar^2 e^2} E_p^2$$

or the density ρ, in kilograms per cubic meter:

$$\rho = \frac{\epsilon_0 m^* M_C}{4 N_A \hbar^2 e^2} E_p^2 \qquad [5.1]$$

where M_C stands for the carbon molar mass and N_A is the Avogadro constant.

Let us note that hydrogen and oxygen are also at least present in the epoxide structure, and contribute for, respectively, 1 and 6 valence electrons:

$$\rho = \frac{\epsilon_0 m^* M_C}{12 N_A \hbar^2 e^2} \frac{11x_C + 15x_O + 1}{3x_C + 5x_O + 1} E_p^2 \qquad [5.2]$$

where x_C and x_O are, respectively, the fractions of carbon and oxygen, putting the hydrogen fraction to $x_H = 1 - x_C - x_O$.

Although equation [5.2] is rigorous, the hydrogen and oxygen contents x_H and x_O in the sample are not necessarily known, but this expression is relatively stable as the composition varies, as shown in Figure 5.1. Furthermore, for a given polymer used as the matrix, we can consider that x_H and x_O are the same in the interphase and bulk. Thus, even when considering that the sample is purely made of carbon, the comparison between the interphase and bulk matrix using equation [5.1] remains quantitatively true.

Three types of nanofillers were tested: (i) commercially available graphite nanoplatelets (GNP) provided by KNANO Science Inc., (ii) graphite oxides (GO), and (iii) a hybrid structure made of carbon nanotubes (CNT) grown on the GNP. The graphite oxides were processed using the modified Hummers method, as described in Liu *et al.* (2014). The CNT were grown on the GNP surface using a chemical vapor deposition (CVD) method, with acetylene as the carbon source and ferrocene as the catalyst precursor, as described in Zhao and Bai (2015).

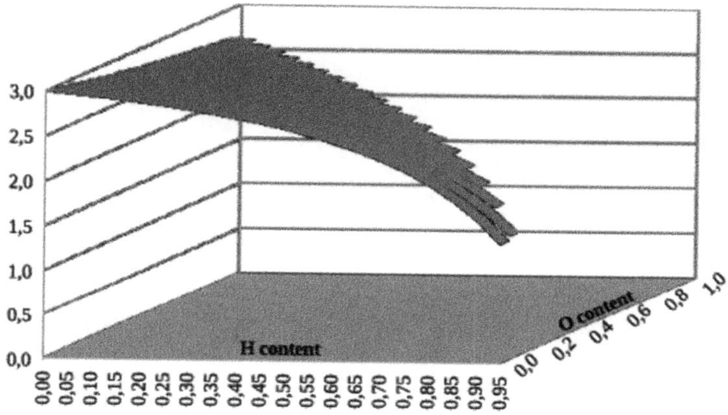

Figure 5.1. *Evolution of the* $\frac{11x_C + 15x_O + 1}{3x_C + 5x_O + 1}$ *ratio used in [5.2], which takes into account the influence of the composition. It shows a great stability with the composition variation safe for a high hydrogen content (close to* 1*), a case which is not often met. For a color version of this figure, see www.iste.co.uk/bai/nanocomposites.zip*

The three types of composites were prepared with a constant composition of 4.0 wt% fillers by mixing 0.8 g of nanofiller with 14.4 g of epoxide in a three-roll mill for 30 minutes, and then manually adding and mixing 4.8 g of a curing agent until a homogeneous suspension was obtained. The sample was then degassed for 60 minutes at room temperature in a vacuum oven, then put in a oven at 60° for 15 hours, and at last annealed at 100° for 90 minutes in order to resume the cross-linking reaction.

The TEM lamellae were ion-milled and tested in a FEI Helios 660 Dual beam, a SEM with a FIB with a Ga^+ source used in order to mill the sample in the wanted area and with the desired geometry. The microscope ion beam resolution is around 4 nm. In our milling conditions, the ion beam current was progressively decreased from 21 to 2.5 nA as the following steps were performed: (i) a 2 μm thick protective layer was deposited on the targeted area where at least a filler and an interface were located, (ii) the sample section, perpendicular to the bulk surface, was milled down to a thickness of approximately 1 μm, (iii) the lamella was extracted and transferred to a TEM grid where it was soldered, (iv) the ultimate fine thinning step was carried out in the desired area of the lamella, using a 200 pA beam current, until it became semi-transparent, and (v) the amorphousized layer was removed using milling conditions from 5 kV and 23 pA to 1 kV and 28 pA (see Figure 5.2).

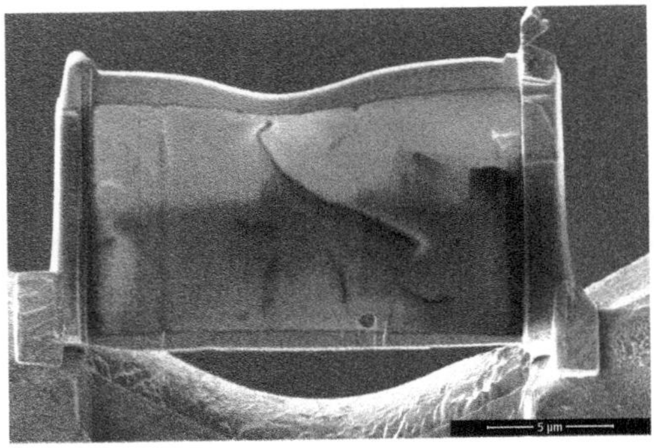

Figure 5.2. *SEM image of the final TEM lamella after the final stage of ion milling*

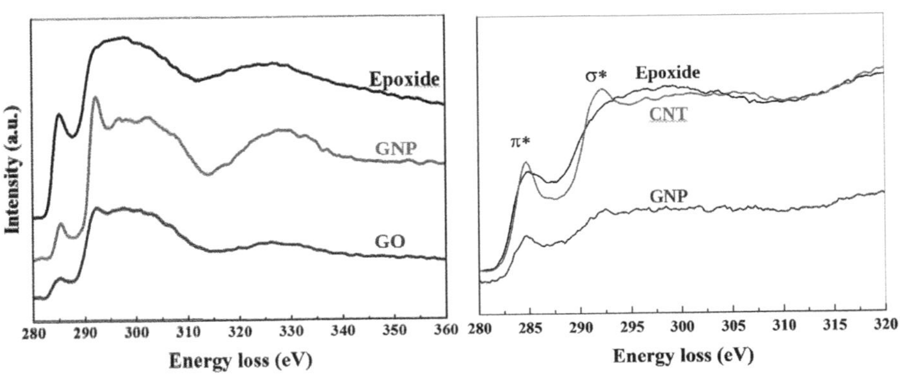

Figure 5.3. *Carbon K-edge spectra of the various samples. (Left) Untreated GNP compared with GO, in epoxide. (Right) CNT grown on GNP in epoxide. For a color version of this figure, see www.iste.co.uk/bai/nanocomposites.zip*

5.2.2. *Results*

The carbon K-edge study confirmed that the different phases observed in imaging mode had the proper chemical structure, as shown in Figure 5.3. The $1s \rightarrow \pi^*$ transition is observed in the first peak at 284 eV, whereas the $1s \rightarrow \sigma^*$ is observed at 291 eV. The difference in height between these peaks is related to the sp^2/sp^3 bonding ratio in the material (Hamon *et al.* 2004), but is also related to observation conditions and more specifically angular effects, as explained in Schattschneider *et al.* (2005). As suspected, the typical graphite fine structure above 291 eV is clearly visible in the untreated GNP and CNT zone, whereas the epoxide fine structure is typical of an amorphous carbon structure. Nevertheless, we observe that the fine structure in the cases of GO and GNP with grown CNT shows a noticeable alteration from the untreated GNP. A reason lies in the local thickness of the sample, since the nanofillers are much harder to ion-mill, and we observed a significant thickness difference between the epoxide domain and the nanofillers.

Low-loss spectra were also acquired in order to measure the local density of the material. In order to have a good comparability of the different phases, the spectra were acquired along a linescan, as shown in the STEM image in the corner of the graph. Two types of linescan were used. The first one was perpendicular to the interface, in order to measure the variability of the plasmon peak position and have the interphase thickness (see Figure 5.4, top). The other was parallel to the interface but in each of the phases (filler, interphase epoxide, bulk epoxide), in order to have the dispersion of the results in one given domain (see Figure 5.4, bottom).

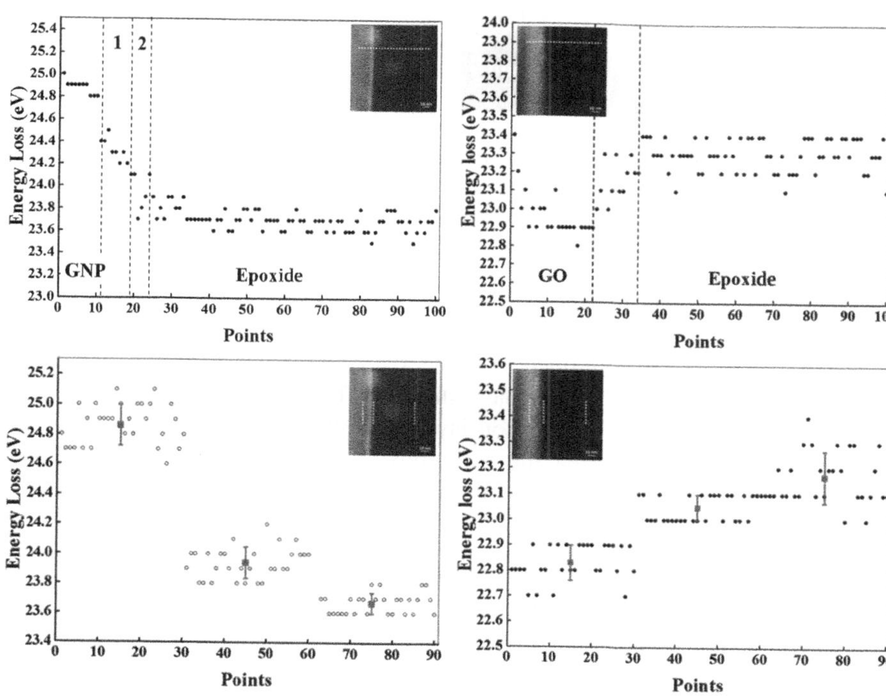

Figure 5.4. *Plasmon peaks measured across different directions, as shown in the STEM image in the corner. Distance between two scanpoints:* 1 *nm. (Left) Untreated GNP in epoxide. (Right) GO in epoxide. For a color version of this figure, see www.iste.co.uk/bai/nanocomposites.zip*

Figure 5.5 shows the plasmon peaks measured at different positions in a hybrid composite made of CNT grown on GNP. Spots 1, 3 and 4 correspond to a CNT, whereas spot 2 corresponds to the epoxide matrix, and spots 5 and 6 to the GNP. The plasmon peak positions are consistent with the spot location in the image.

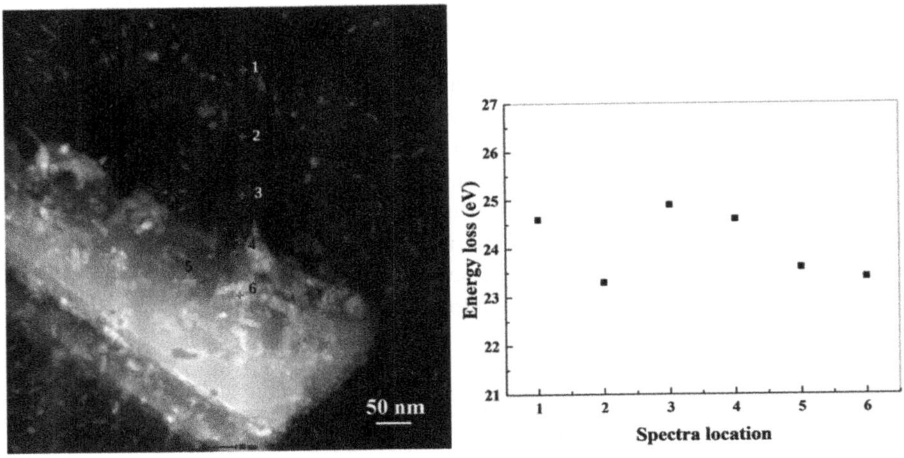

Figure 5.5. *(Left) TEM view of a zone with CNT grown on GNP (light gray) in epoxide (dark gray). (Right) Plasmon peak positions in the six spots shown on the image. For a color version of this figure, see www.iste.co.uk/bai/nanocomposites.zip*

5.2.3. *Interpretation*

The K-edge spectra confirmed that the oxidation process of the nanofillers generates a modification of the chemical structure. The plasmon peak measurement confirmed the existence of a transition layer, the interphase, where the density is neither the one of the filler nor the bulk epoxide, and whose thickness could be measured with the cross-sectional scanline: from 12 to 12.5 nm for GNP/epoxide and GO/epoxide.

In the GO/epoxide case, we observed a drop in the densities. First, the density of GO is decreased compared to GNP, due to the increase in the interplane distance of the crystal structure. Then, at the interface, the oxygen-containing groups introduced during the oxidation process react with the epoxide curing agent and the interface is blurred. Furthermore, in section 5.2.1 it was assumed that the chemical composition, if not precisely known, was constant in the bulk epoxide and in the interphase. It is certainly the case when no chemical treatment has been used, but in the case of GO/epoxide, the oxygen content is significantly increased, as confirmed by energy-dispersive X-ray (EDX) spectroscopy (see Figure 5.7). Thus, the density calculation contains a lot of uncertainty. Nevertheless, the in-depth oxidation of graphite dramatically modifies the filler properties, by suppressing C–C bonds and forming new C–O bonds. Such an alteration of the properties is not necessarily favorable in terms of mechanical properties.

We cannot discuss an interphase thickness in the case of the CNT-GNP hybrid, since the roughness of the interface prevents from any notion of a sharp change of domain.

5.3. *In situ* tensile test using scanning electron microscopy

Tensile tests were performed on a rectangular shaped sample of approximately $13 \times 4 \times 1.5$ μm dimension, half in the CF zone and the other half in the epoxide matrix. The fiber part was connected to a copper massif, whereas the tensile load was externally applied on the epoxide side by the microscope nanomanipulator. Four types of CF were tested, and each was tested five to eight times. Depending on the sample, the length of the epoxide rectangular area ranged from 6.5 to 7 μm.

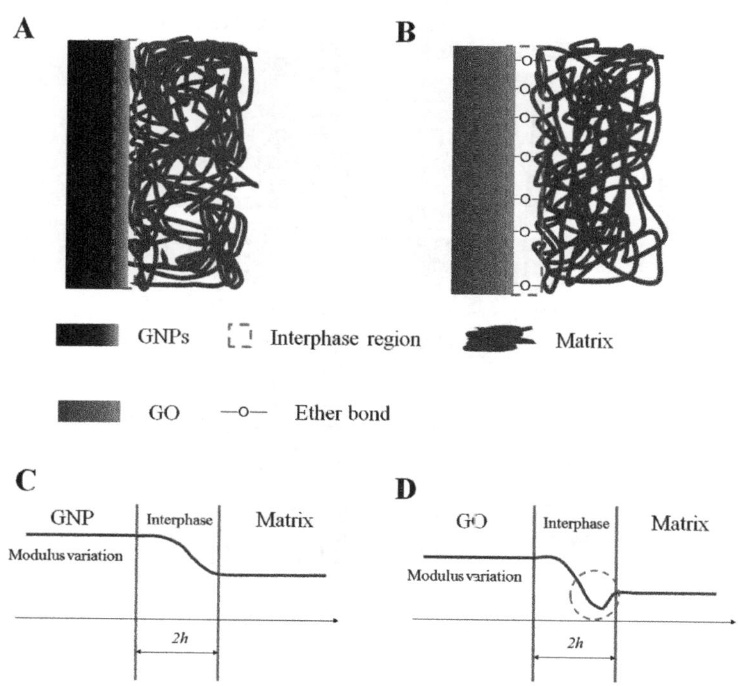

Figure 5.6. *Sketch representing the composite interface and interphase, and the corresponding modulus variation using Delale and Erdogan (1988). (A,C) GNP/epoxide composite (B,D) GO/epoxide. For a color version of this figure, see www.iste.co.uk/bai/nanocomposites.zip*

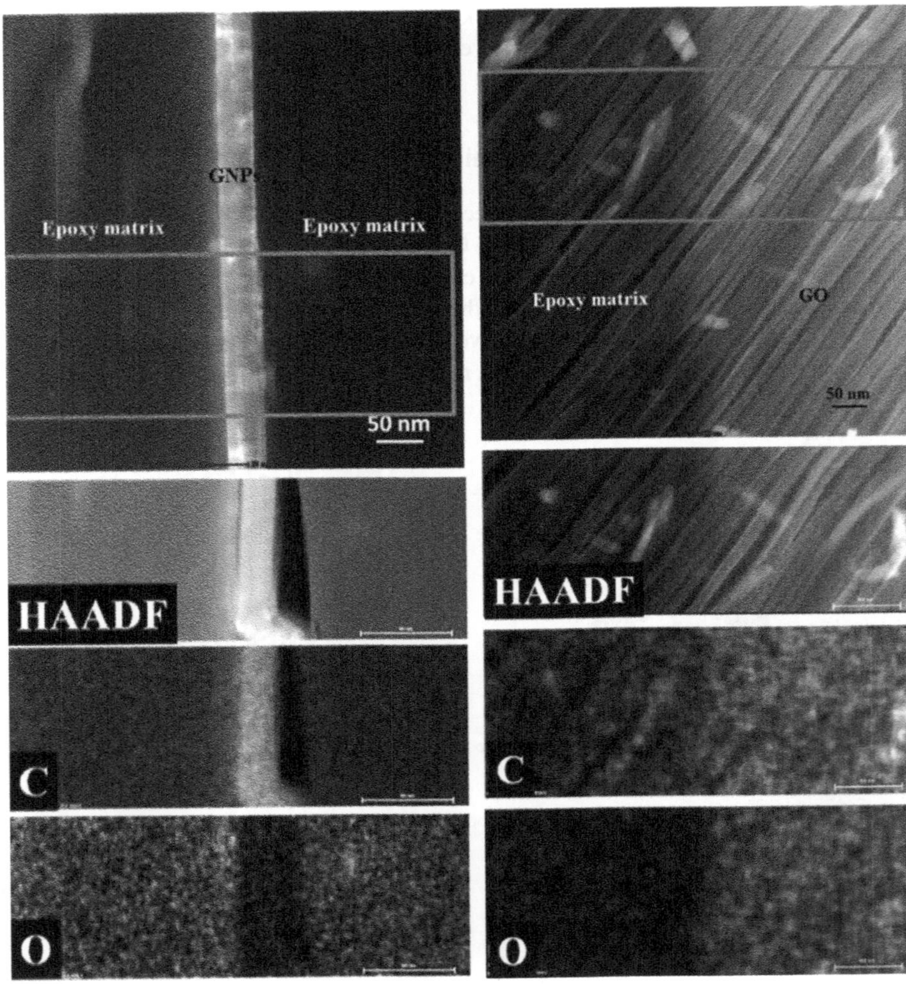

Figure 5.7. *TEM imaging and EDX analysis of (right) GNP epoxide, and (left) GO/epoxide composites exhibiting the carbon and oxygen contents, which confirm the in-depth oxidation process of GO. For a color version of this figure, see www.iste.co.uk/bai/nanocomposites.zip*

5.3.1. *Experimental set-up*

The samples were ion milled and tested in the same FEI Helios 660 Dual beam as described in section 5.2.1. The ion beam resolution is around 4 nm, and the ion beam current was progressively decreased from 21 to 2.5 nA as the following steps were performed (see Figure 5.8): (i) circular-shaped CF sections were located at the surface of the bulk material by SEM imaging; (ii) a protective platinum layer with a rectangular shape ($20 \times 2 \times 1$ μm) was deposited across the selected fiber-epoxide interface; (iii) the sample was milled down to a thickness of 1 μm and a width of 8 μm; (iv) the lamella was lifted out by the nanomanipulator and soldered to a copper grid by Pt deposition; (v) at the fiber–copper interface, two quarter-circles were milled in order to decrease the stress concentration; (vi) the nanomanipulator was soldered to the lamella by Pt deposition. It is noteworthy that the sample curvatures were designed on the part of the sample with the higher strength, so that the stress concentration appearing locally next to these discontinuities had no influence on our measurement, since it was much smaller than the rupture criterium of copper fiber or CF.

Four types of CF treatments were studied: (i) commercially available raw CF, which included a sizing layer on the surface, (ii) desized CF, obtained after the removal of the sizing layer, (iii) hybrids made up of carbon nanotubes (CNT) grown on the CF surface, (iv) oxidized hybrids made up of the previous CNT grown on the CF surface and then oxidized. For this study, we chose a T700 grade CF, with a diameter of around 7 μm, provided by Toray Carbon Fibers America, Inc. The desized CF were obtained by heating the as-received CF up to a temperature of 650° in an Ar atmosphere for several minutes (Zhang 2012). The CNT were grown on CF using a CVD method in a furnace at 650° under an Ar and H flow, with respective rates of 1.2 and 0.3 L.min^{-1}. After a stabilization period, a xylene solution containing 0.1 g.mL^{-1} ferrocene was injected. After 3 minutes of growth, the furnace was cooled down to room temperature under an Ar protective atmosphere (Bai 2010). The fourth sample was obtained from the previous CNT-CF one by heating at 400° and stabilizing for 4 minutes in air.

5.3.2. *Results*

There was no direct indication of the load applied to the sample during the tensile test: a step-by-step displacement of 200 nm was imposed to the nanomanipulator, and the stress field was inferred from the global strain field in the sample. In particular, the transverse length in the epoxide domain was analyzed and compared with macroscopic tensile tests performed previously on a pure epoxide sample. In order to strengthen the results, observations were made using both the electron imaging mode and the ion mode, with two different viewing angles. Both curves overlapped almost everywhere, as shown in Figure 5.9.

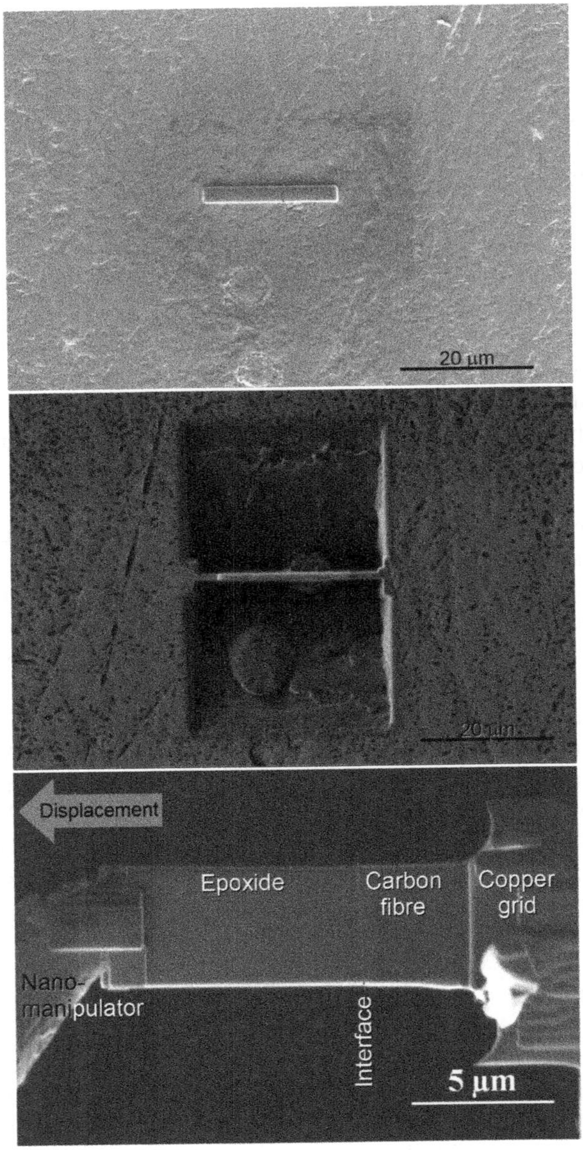

Figure 5.8. *Different stages of the sample preparation. (Top) A protective platinum layer is deposited over a rectangular area where a carbon fiber (red circle) was located. (Middle) The sample – whose plane is perpendicular to the bulk surface – is ion milled. Carbon fibers, much harder than epoxide, can be spotted in the excavated zone. (Bottom) Final experimental setup*

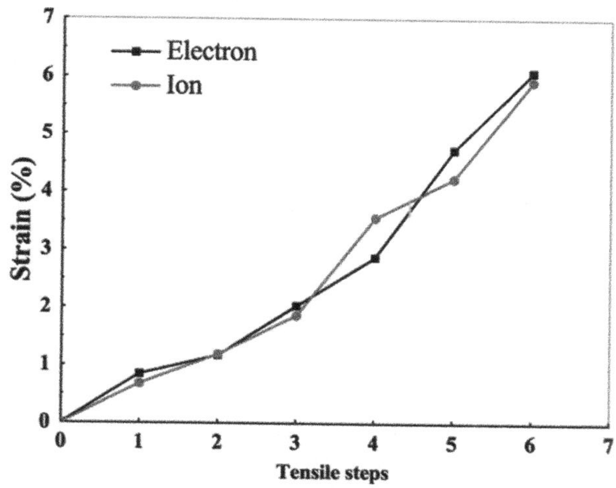

Figure 5.9. *Comparison of the transverse strain–displacement curves measured in the epoxide from the electron and ion imaging modes. Example obtained on a raw, commercially available, carbon fiber. For a color version of this figure, see www.iste.co.uk/bai/nanocomposites.zip*

In some samples, the rupture was observed inside the epoxide area, exhibiting a higher strength at the interface. For others, it was observed in the interface zone, emphasizing that the surface modification at the interface fulfilled its function. Figure 5.10 (left) shows the displacement–strain curves obtained with all samples, and Figure 5.10 (right) shows the ultimate strength measured during these same tensile tests.

We observe that the samples with CNT-CF hybrids give significantly better results than raw CF. Moreover, the oxidation process of the CNT-CF hybrids, which cumulates both types of surface treatment strategies mentioned in the introduction, gives a higher strength than the non-oxidized hybrids. Nevertheless, there are discrepancies in the results that must be carefully analyzed.

5.3.3. *Interpretation*

For the cases giving the better results, we observe that the strength compares with the ultimate strength of the bulk epoxide. Actually, for the two highest values measured in the oxidized hybrid CNT-CF, the rupture was observed in the epoxide domain, quite far from the interface. The surface treatment shows its limit: the tensile stress in the bulk epoxide cannot be decreased through a structural modification, and is the weak point in the composite.

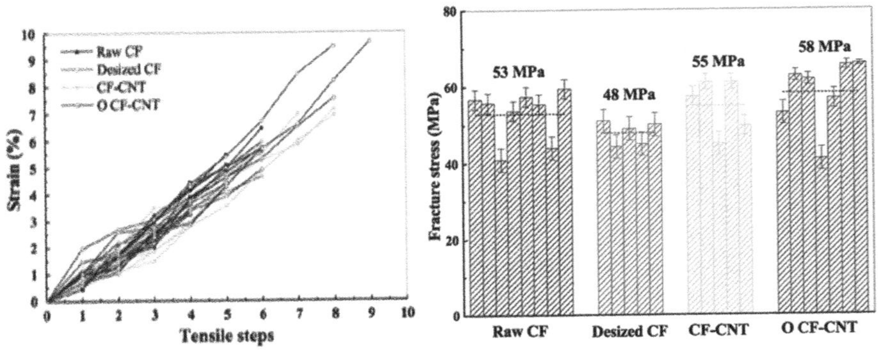

Figure 5.10. *Tensile* in situ *tests on the four types of treated carbon fibers (CF): raw CF, CF after the removal of their sizing layers, hybrids made up of carbon nanotubes (CNT) grown on CF, oxidized CNT-CF hybrids. (Left) Strain–stress curves for all* 25 *tests; (right) corresponding ultimate strength. For a color version of this figure, see www.iste.co.uk/bai/nanocomposites.zip*

The second point of interest concerns the measures giving the lowest results. More particularly, we observe that these results, observed with the modified CF, compare with the case of the desized CF, and lead to two notes: (i) the desized CF present the lowest values, but also the one with the smallest dispersion, with a strength around 48 GPa; (ii) for the other samples, it implies a problem of repeatability in the surface treatment. The dispersion in the results is an indirect measurement of the treatment homogeneity. For instance, concerning the raw CF, the sizing layer is probably not homogeneous along the length of the fiber, with the added fact that they have probably been bundled up before being permeated with the sizing agent, and CF positioned in the bundle center are less treated than those at the periphery.

5.3.4. *Perspectives*

Because of its geometry, this tensile testing gives an account of phenomena that are in the direction perpendicular to the interface. Other phenomena may occur in the plane of the interface. In that case, the choice of the surface modification could lead to different results, as the physical model behind is different: for oxidized fillers, it is not the elongation of the bond which is at stake, but its angular distortion. *In situ* shear tests would produce the desired data, although these results have already been tackled by the authors on macroscopic samples (Liu *et al.* 2016).

Another point is the result dispersion that highlights the well-known difficulty of process reproducibility, especially in a mass production perspective (relatively speaking). It is only one of the many parameters that must be controlled along with the filler distribution, and it is an endless source of improvement.

5.4. Conclusion

We carried out different local investigations on nanofiller/epoxide interfaces made possible by the latest microscopy techniques. EEL spectroscopy confirmed the existence of an interphase by measuring the local density across the filler/epoxide interface, which was around 12 nm thick, both in the case of GNP and GO fillers. It also led to the conclusion that the chemical treatment, namely the oxidation of the GNP to obtain GO, generates a modification of the chemical composition, and leads to a drop in the density, both in the filler and the interphase.

The effect of the filler treatment was also investigated through *in situ* tensile testing, where an isolated interface could be tested. Unremarkably, the filler that cumulated both types of surface modifications (CNT grown on the surface and oxidation) gave the best ultimate tensile strength. It also showed the limit of the process, since in that case, the fracture was observed in the bulk epoxide matrix, displacing the weak point of the composite in the bulk matrix.

These measures can be improved by a better energy resolution in the case of EEL spectroscopy supported with a consistent model, which could relate the structure modification and the local alteration of the material properties. Moreover, other *in situ* tensile tests with information applied on the load could strengthen the knowledge of the mechanical behavior in the interface zone, depending on the filler modification.

5.5. Acknowledgments

The results presented here were obtained from the PhD thesis of Yu Liu, who was granted a CSC scholarship. The authors wish to thank Jean-François Gérard, Isabelle Royaud, Gilles Lubineau, Lionel Pichon and Gilles Régnier for fruitful discussions.

This work was carried out within the MATMECA consortium and supported by the ANR under contract number ANR-10-EQPX-37. It has benefited from the facilities of the Laboratory MSSMat, CNRS, CentraleSupélec, Université Paris-Saclay, France.

5.6. References

Bai, J. (2000). Procédé de synthèse de nanotubes de carbone sur matériaux micrométriques longs et particulaires. Patent no. EP2370355A2.

Delale, F. and Erdogan, F. (1988). On the mechanical modeling of the interfacial region in bonded half-planes. *Journal of Applied Mechanics*, 55(2), 317–324.

Egerton, R.F. (1986). *Electron Energy-loss Spectroscopy in the Electron Microscope*. Plenum Press, New York.

Ferrari, A.C., Libassi, A., Tanner, B.K., Stolojan, V., Yuan, J., Brown, L.M., Rodil, S.E., Kleinsorge, B., Robertson, J. (2000). Density, sp^3 fraction, and cross-sectional structure of amorphous carbon films determined by X-ray reflectivity and electron energy-loss spectroscopy. *Phys. Rev. B*, 62(16), 11089–11103.

Hamon, A.-L., Verbeeck, J., Schryvers, D., Benedikt, J., Van de Sanden, M.C.M. (2004). ELNES study of carbon K-edge spectra of plasma deposited carbon films. *J. Mater. Chem.*, 14(13), 2030–2035.

Im, H. and Kim, J. (2012). Thermal conductivity of a graphene oxide-carbon nanotube hybrid/epoxy composite. *Carbon*, 50(15), 5429–5440.

Liu, Y., Zhang, Y., Zhang, T., Jiang, Y., Liu, X. (2014). Synthesis, characterization and cytotoxicity of phosphorylcholine oligomer grafted graphene oxide. *Carbon*, 71, 166–175.

Liu, Y., Hamon, A.-L., Bai, J. (2016). Directly measuring interfacial shear strength between Polymethyl methacrylate and graphene nanoplatelets. *Materials Lett.*, 182, 244–247.

Pukánszky, B. (2005). Interfaces and interphases in multicomponent materials: Past, present, future. *European Polymer Journal*, 41(4), 645–662.

Ramanathan, T., Abdala, A.A., Stankovich, S., Dikin, D.A., Herrera-Alonso, M., Piner, R.D., Adamson, D.H., Schniepp, H.C., Chen, X., Ruoff, R.S., Nguyen, S.T., Aksay, I.A., Prud'Homme, R.K., Brinson, L.C. (2008). Functionalized graphene sheets for polymer nanocomposites. *Nature Nanotechnology*, 3, 327.

Saba, J., Magga, Y., He, D., Miomandre, F., Bai, J. (2013). Continuous electrodeposition of polypyrrole on carbon nanotube-carbon fiber hybrids as a protective treatment against nanotube dispersion. *Carbon*, 51, 20–26.

Schattschneider, P., Souche, C., Franco, H., Jouffrey, B. (2005). Anisotropic relativistic cross sections for inelastic electron scattering, and the magic angle. *Phys. Rev. B*, 72(4), 045142.

Sengupta, R., Bhattacharya, M., Bandyopadhyay, S., Bhowmick, A.K. (2011). A review on the mechanical and electrical properties of graphite and modified graphite reinforced polymer composites. *Progress in Polymer Science*, 36(5), 638–670.

Wu, Q., Li, M., Gu, Y., Li, Y., Zhang, Z. (2014). Nano-analysis on the structure and chemical composition of the interphase region in carbon fiber composite. *Composites Part A: Applied Science and Manufacturing*, 56, 143–149.

Zhang, J. (2012). Different surface treatments of carbon fibers and their influence on the interfacial properties of carbon fiber/epoxy composites. PhD Thesis, École Centrale, Paris.

Zhao, H. and Bai, J. (2015). Highly-sensitive piezo-resistive graphite nanoplatelet-carbon nanotube hybrids/polydimethylsilicone composites with improved conductive network constructions. *ACS Applied Materials & Interfaces*, 7(18), 9652–9659.

6

Multiscale Modeling of Graphene-polymer Nanocomposites with Tunneling Effect

Xiaoxin LU[1,2,3], Julien YVONNET[2], Fabrice DETREZ[2] and Jinbo BAI[3]

[1] *Shenzhen Institutes of Advanced Technology, Chinese Academy of Sciences, China*
[2] *MSME, Gustave Eiffel University, Marne-la-Vallée, France*
[3] *CentraleSupélec, Paris-Saclay University, Gif-sur-Yvette, France*

6.1. Introduction

The introduction of graphene-based nanomaterials has prompted the development of flexible nanocomposites for emerging applications in various fields, including energy conversion (Britnell *et al.* 2013), energy storage (El-Kady and Kaner 2013), electronic materials (Kim *et al.* 2009), sensors (Mannoor *et al.* 2012) and chemical screening applications (Guo *et al.* 2013). Graphene is among the materials with the highest in-plane electric conductivity (Novoselov *et al.* 2004), and its incorporation in a polymer matrix to increase the electric conductivity of almost insulating polymers is of high importance for materials-by-design. Numerous experimental and theoretical studies have reported that dispersing two-dimensional fillers such as graphene sheets in a polymer matrix can significantly improve the electric properties of the resulting composites (Liang *et al.* 2009, Kim and Macosko 2009, Hicks *et al.* 2009, Qi *et al.* 2011, Wang *et al.* 2015, He and Tjiong 2013). They exhibit an increase in several orders of magnitude of the electric conductivity, even at extremely

Nanocomposites,
coordinated by Jinbo BAI.

low volume fractions of graphene sheets (Fan *et al.* 2012, Tkalya *et al.* 2014, Gao *et al.* 2014, Trionfi *et al.* 2009, Du *et al.* 2005), denoting the percolation phenomenon. This phenomenon can be explained by the formation of percolating paths between graphene sheets through the matrix. The percolation threshold of a random assembly of widthless discs has been evaluated in last century for the permeability problem (Charlaix 1986), dominated by the radius and number of discs. However, the introduction of nanofillers brings more complexity to the study of connectivity percolation of the composite.

Experimental works have evidenced that electrical properties of graphene or nanotube-reinforced nanocomposites are significantly dependent on the microstructure parameters, such as the size (Bryning *et al.* 2005) and the orientation of fillers (Yousefi *et al.* 2012, Du *et al.* 2005), as well as the inherent characteristics of the polymer matrix (Bauhofer and Kovacs 2009, Haggenmueller *et al.* 2007).

For design purposes, numerical models of conduction at the scale of a representative volume element (RVE) containing a significant number of graphene sheets are required to fully understand the conditions for percolation and determining optimal configurations and microstructural parameters to increase the performances of these materials. Nowadays, complete ab initio or atomistic simulations including electric conductivity in large systems like polymer–graphene reinforced composites are not feasible, and classical homogenization methods or Monte Carlo techniques (Bauhofer and Kovacs 2009, Castaneda and Willis 1995, Fan *et al.* 2015, Xia *et al.* 2017, Wang *et al.* 2014, Grimaldi and Balberg 2006, Otten and van der Schoot 2009) are unable to explain the nonlinear effects and low percolation thresholds in graphene–polymer nanocomposites. Simulations of larger systems require continuum descriptions of fields and related numerical methods.

In this chapter, we present a multiscale nonlinear modeling framework for graphene–polymer composites based on RVE calculations (Lu 2017, Lu *et al.* 2018, Lu *et al.* 2019). We first present a continuum framework to model the nonlinear electrical conduction at the scale of a RVE. The advantage of such model is that the microstructural parameters (volume fraction, orientation of platelets, etc.) can be changed to analyze their sensitivity with respect to the effective electrical conduction properties. The tunneling effect is taken into account to describe the nonlinear effects and low percolation thresholds observed in the literature. Then, the same model is used in structural applications. As the effective behavior of the RVE is strongly nonlinear, it is simply not possible to use an identified empirical model at the macroscopic scale. Then, a numerical two-scale approach is proposed, where the nonlinear behavior is identified by recent machine learning methods. Finally, the effect of the interphases between graphene and the polymer matrix is investigated to study its impact on the effective piezoresistive (electromechanical) behavior of the composite. A molecular dynamics (MD) approach is proposed to identify the nonlinear cohesive interfacial model.

6.2. Modeling of effective electric nonlinear behavior in graphene–polymer nanocomposites

6.2.1. *Tunneling effect*

Tunneling electric conductivity is a quantum phenomenon that allows electric conductivity across small isolating barriers, like thin polymer interphases between two highly conducting fillers. It has been shown that electric tunneling effect plays an important role in explaining very low percolation thresholds, as well as nonlinear electric conduction (Oskouyi *et al.* 2014, Sheng 1980) in such materials, especially when the characteristic distances between graphene sheets reduce to the order of nanometers. In the works of Otten and van der Schoot (2011) and Ambrosetti *et al.* (2010), the tunneling effect has been taken into account to analyze the percolation threshold of nanocomposites with polydisperse nanofillers, where the former focuses on carbon nanotubes and the latter extends to various particle shapes.

In a first approximation, the polymer–graphene nanocomposite can be seen as a juxtaposition of elementary graphene–polymer–graphene units (see Figure 6.1(a)). In order to understand the nonlinear electrical behavior of the nanocomposite, we first focus on the conduction mechanism of this unit. When the distance between two graphene sheets is nanoscaled, the polymer matrix can be seen as a dielectric film between two graphene electrodes. There is a strong analogy with a metal–insulator–metal device with tunnel junction. It is known that there are two types of conduction mechanisms in dielectric films (Hamann *et al.* 1988), which are interface-limited conduction mechanisms and bulk-limited conduction mechanism, respectively. The former depends on the electric properties at the polymer–graphene interface, while the latter depends on the electric properties of the polymer. In our work, we focus on solving the general problem and neglect the bulk-limited conduction mechanism, except Ohm's law in varying polymers.

There are several interface-limited conduction mechanisms (Chiu 2014): (1) Schottky emission; (2) thermionic-field emission; (3) Fowler–Nordheim tunneling; (4) direct tunneling. Among them, the former two are only applied at high temperatures, when the electrons can get enough thermal energy to overcome the barrier so that the electronic current can flow in the conduction band. The latter two occur when the barrier is thin enough to permit its penetration by the quantum tunnel (Simmons 1971). To simplify the problem, we assume that the system is studied at low temperatures so that the thermal current can be neglected, and restrict the electron transportation between electrodes to the tunneling effect.

Figure 6.1 depicts the evolution of conduction band edge energy, E_c, which is at an equilibrium state across the graphene–polymer–graphene device in Figure 6.1(a).

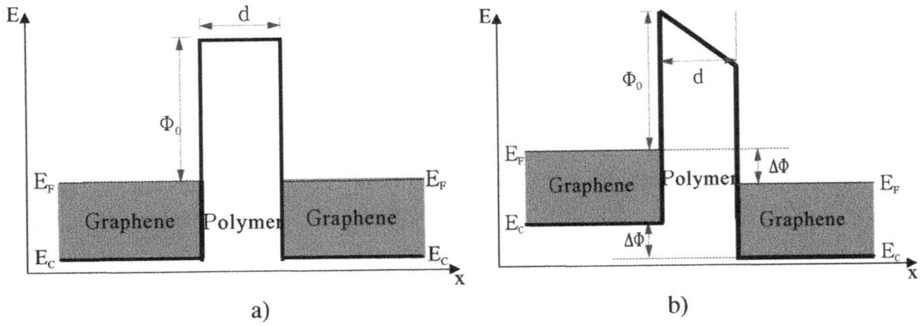

Figure 6.1. *Band diagrams of graphene–polymer–polymer unit a) at equilibrium $V = 0$ b) with applied voltage V. The gray part indicates filling of energy levels. E_c is the conduction band edge energy and E_F is the fermi level in graphene (for reasons of clarity in $E_F \neq E_c$, although it is the case in the graphene monolayer)*

For such a 1D model of two close graphene sheets separated by a polymer matrix, the electric constitutive law is expressed in the polymer in the form

$$I = \mathcal{G}\,(E, d)\,E \tag{6.1}$$

where I denotes the electric current, E is the electric field and $\mathcal{G}\,(E, d)$ is a nonlinear function of the electric field E and the distance between the two graphene sheets d. An explicit formula for the electric tunneling effect through a potential square barrier was first derived by Simmons (1963) as:

$$\begin{aligned} \mathcal{G}\,(E, d) = \; & \frac{2.2e^3E^2}{8\pi h\Phi_0} \exp(-\frac{8\pi}{2.96heE}(2m)^{\frac{1}{2}}\Phi_0^{\frac{3}{2}})\ldots \\ & +[3 \cdot \frac{(2m\Phi_0)^{\frac{1}{2}}}{2d}](e/h)^2 Ed \exp[-(\frac{4\pi d}{h})(2m\Phi_0)^{\frac{1}{2}}] \end{aligned} \tag{6.2}$$

Here, the tunneling current I is a nonlinear function of electric field E and depends on three parameters: the height Φ_0, width d of the barrier and the effective mass m of the electrons in graphene. This expression is obtained under the assumptions that the temperature is very low (i.e. $T \approx 0$ K); the electric field E varies slowly over the width of the potential barrier (i.e. $E \approx V/d$, and the rectangular barrier becomes trapezoidal under the effect of the electric field (see Figure 6.1(b))), and the trapezoidal barrier is approximated by the rectangular barrier with the same average height. This theory has been used and validated experimentally in the metal–insulator–semiconductor (Maity *et al.* 2016) and metal–insulator–metal capacitor field (Krishnan *et al.* 2008). Note that the current only comes from the electrons that cross the potential barrier. In Simmons' model, the leakage current is not taken into account. This can be covered in the 3D extension of the conductivity model (see section 6.2.2). However, since we consider the bulk current in addition to the tunneling current, the leakage current will not affect the effective current in the composite.

6.2.2. *Nonlinear electrical conduction model at the RVE scale*

In this section, we describe a continuum RVE model for the graphene–polymer nanocomposite. At this scale, the graphene sheets are explicitly described at the scale of a quite large cluster of graphene nano platelets (see Figure 6.2) composed of a polymer matrix and randomly distributed and oriented graphene sheets, modeled as highly conducting imperfect interfaces (Yvonnet *et al.* 2008).

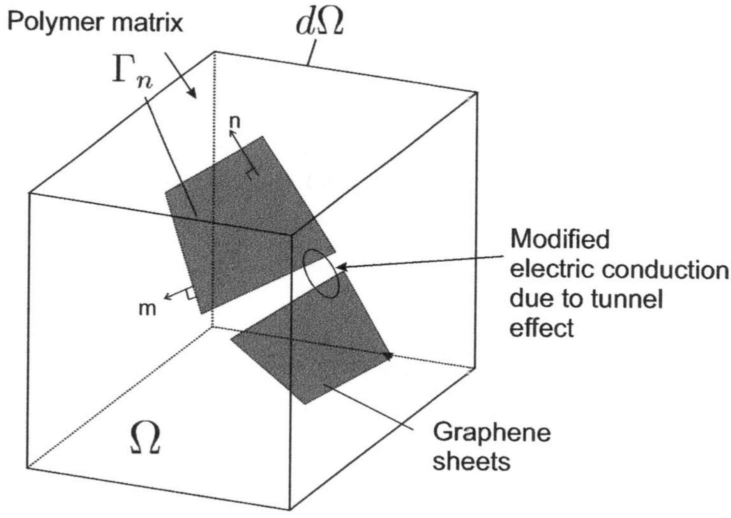

Figure 6.2. *RVE model of the graphene-reinforced composite*

The electric tunneling effect between graphene sheets described below is taken into account in the continuum model.

In this model, the total electric power of the system, $\overline{W}$, is defined by

$$2\overline{W} = \int_{\Omega} \omega^b(\mathbf{x})d\Omega + \int_{\Gamma} \omega^s(\mathbf{x})d\Gamma, \qquad [6.3]$$

where $\Omega \subset \mathbb{R}^3$ is the domain defining the RVE, Γ denotes the surfaces associated with graphene collectively and the density functions ω^b and ω^s are expressed by

$$\omega^b(\mathbf{x}) = j_i(\mathbf{x})E_i(\mathbf{x}), \quad \omega^s(\mathbf{x}) = j_i^s(\mathbf{x})E_i^s(\mathbf{x}), \qquad [6.4]$$

where $E_i(\mathbf{x}) = -\nabla_i\phi(\mathbf{x}) = -\phi_{,i}(\mathbf{x})$ is the electric field, $\mathbf{j}(\mathbf{x})$ is the current density vector and $\phi(\mathbf{x})$ is the electric potential, where:

$$j_i = \left[k_p^0\right]_{ij} E_j \qquad [6.5]$$

with $\mathbf{k}_p^0$ as the electric conductivity tensor of the polymer matrix when neglecting the tunneling effect. In the above, the superscript s denotes surface quantities, for example, $\mathbf{j}^s$ is the surface current density. The surface electric field is defined with respect to its bulk counterpart as: $E_i^s = P_{ij}E_j = -\nabla_i^s\phi = -P_{ij}\nabla_i\phi$ with $P_{ij}(\mathbf{x}) = \delta_{ij} - n_i(\mathbf{x})n_j(\mathbf{x})$ as a projector operator characterizing the projection of a vector along the tangent plane to Γ at a point $\mathbf{x} \in \Gamma$, and $\mathbf{n}$ is the unit normal vector to Γ. The local constitutive relationships relating $\mathbf{j}$ and $\mathbf{j}^s$ with $\mathbf{E}$ are defined in the graphene sheets by

$$j_i^s(\mathbf{x}) = k_{ij}^s E_j^s, \qquad [6.6]$$

where $\mathbf{k}^s$ denotes the surface electric conductivity of graphene and is dependent on the thickness t through:

$$k_{ij}^s = tS_{ij}, \quad S_{ij} = [k^g]_{ij} - \frac{[k^g]_{ik}n_k[k^g]_{jk}n_k}{[k^g]_{kl}n_k n_l}. \qquad [6.7]$$

In [6.7], $\mathbf{k}^g$ denotes the second-order electric conductivity tensor of the bulk graphite and $\mathbf{n}$ is the normal vector to the graphene sheet (for more details, see Lu *et al.* (2018), Yvonnet *et al.* (2008)).

$$j_i = \begin{cases} \left[k_p^0\right]_{ij} E_j & \text{if } d(\mathbf{x}) \geq d_{cut}, \\ \mathcal{G}(\mathbf{E}, d)\frac{E_i}{|\mathbf{E}|} & \text{if } d(\mathbf{x}) < d_{cut}, \end{cases} \qquad [6.8]$$

where d_{cut} is a cut-off distance above which the tunneling effect can be neglected and $\mathcal{G}$ is a 3D extension of [6.2] defined by:

$$\mathcal{G}(\mathbf{E}, d) = \frac{2.2e^3 |\mathbf{E}|^2}{8\pi h\Phi_0} \exp\left(-\frac{8\pi}{2.96he|\mathbf{E}|}(2m)^{\frac{1}{2}}\Phi_0^{\frac{3}{2}}\right)$$

$$+[3 \cdot \frac{(2m\Phi_0)^{\frac{1}{2}}}{2}](e/h)^2 |\mathbf{E}| \exp[-(\frac{4\pi d}{h})(2m\Phi_0)^{\frac{1}{2}}] \quad [6.9]$$

where Φ_0 is the energy barrier height that the electrons cross and h, e and m denote Plank's constant, the charge of an electron and a material parameter, and d is a distance function defined in 3D (see more details in Lu *et al.* (2018)).

Minimizing [6.3] with respect to the displacement field, and using equations [6.5]–[6.8], we obtain the weak form, which can be solved by using the finite element method (see more details in Lu *et al.* (2018)):

$$\int_{\Omega} j_i(\phi) \cdot \nabla(\delta\phi) d\Omega - \int_{\Gamma} P_{ij} \nabla_j \phi k^s_{ik} P_{kl} \nabla_l(\delta\phi) d\Gamma = 0, \quad [6.10]$$

where $\delta\phi \in H^1(\Omega)$, $\delta\phi = 0$ over $\partial\Omega$ and $\phi \in H^1(\Omega)$, ϕ satisfying the periodic boundary conditions over $\partial\Omega$

$$\phi(\mathbf{x}) = -\overline{E}_i x_i + \tilde{\phi}(\mathbf{x}) \qquad \text{on } \partial\Omega \quad [6.11]$$

and where $\tilde{\phi}(\mathbf{x})$ is a periodic function over Ω, such as $\left\langle \tilde{\phi}(\mathbf{x}) \right\rangle = 0$.

The effective electric conductivity tensor $\overline{\mathbf{k}}$ is defined as:

$$\overline{k}_{ij}(\overline{\mathbf{E}}) = \frac{\partial \overline{J}_i(\overline{\mathbf{E}})}{\partial \overline{E}_j}, \quad [6.12]$$

where $\overline{\mathbf{J}}$ is the effective current density whose components are expressed by:

$$\overline{J}_i = \frac{1}{V} \left(\int_{\Omega} j_i(\mathbf{x}) d\Omega + \int_{\Gamma} j^s_i(\mathbf{x}) d\Gamma \right), \quad [6.13]$$

and $\overline{\mathbf{E}}$ is the effective electric field given by:

$$\overline{E}_i = \frac{1}{V} \int_{\Omega} E_i(\mathbf{x}) d\Omega \quad [6.14]$$

6.3. Numerical simulations of effective electric conductivity

6.3.1. *Effect of barrier height on the percolation threshold*

Using the above model, we study the effects of barrier height between graphene and various polymers on the nonlinear response of the nanocomposite. In our simulations, we have considered that the percolation threshold corresponds to a sharp variation of the effective conductivity above 10^{-8} S/m. In Figure 6.3, the effective conductivity component $\overline{k}_{11}$ is plotted as a function of the graphene volume fraction for the values $\Phi_0 = 0.17$ eV, 0.3 eV, 1.0 eV and without tunneling effect. The aspect ratio is $\eta = 50$ and applied electric field $\overline{E} = 1.25 \times 10^{-3}$ V/nm. It should be noted that for small barrier height (0.17 eV and 0.3 eV), the computation results of electric conductivity, taking into account the tunneling effect, are much larger than the predictions without tunneling effect, while for $\Phi_0 = 1.0$ eV, the effective conductivity characteristics exhibit no obvious difference, either with or without tunneling effect.

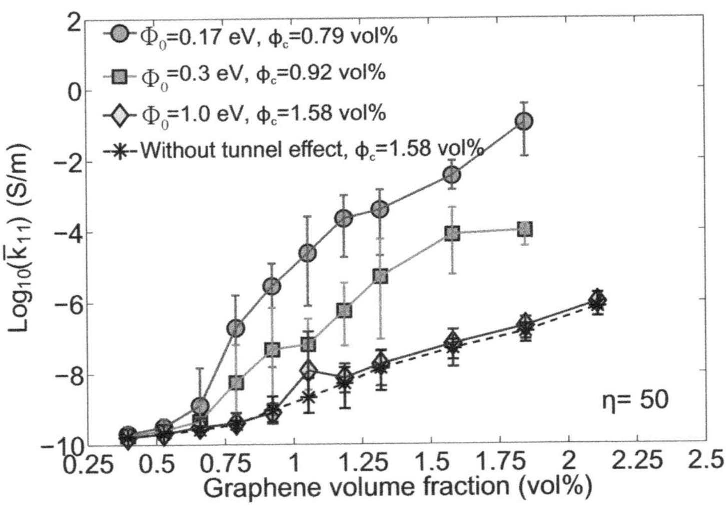

Figure 6.3. *Effective conductivity versus graphene sheets volume fraction for several barrier heights* $\Phi_0, \overline{E} = 1.25 \times 10^{-3}$ *V/nm,* $\eta = 50$. ϕ_c *denotes the percolation threshold. For a color version of this figure, see www.iste.co.uk/bai/nanocomposites.zip*

The percolation thresholds corresponding to $\Phi_0 = 0.17, 0.3, 1.0$ eV are 0.79 vol%, 0.92 vol% and 1.58 vol%, respectively. We can see from these simulations that the lower the barrier height is, the lower the percolation threshold. Experimental results reporting percolation thresholds for different polymer matrices and graphene types can be found in previous studies (Stankovich *et al.* 2006, Jiang *et al.* 2009, Ansari and Giannelis 2009, Liang *et al.* 2009).

6.3.2. *Effect of graphene aspect ratio on the percolation threshold*

Next, we use our numerical model to estimate the percolation threshold f^* of the nanocomposite as a function of the graphene volume fraction and aspect ratio for graphene sheets. In Figure 6.4, the effective conductivity tensor component $\overline{k}_{11}$ is computed for different aspect ratios $\eta = 20, 50$ and 100 as a function of volume fraction.

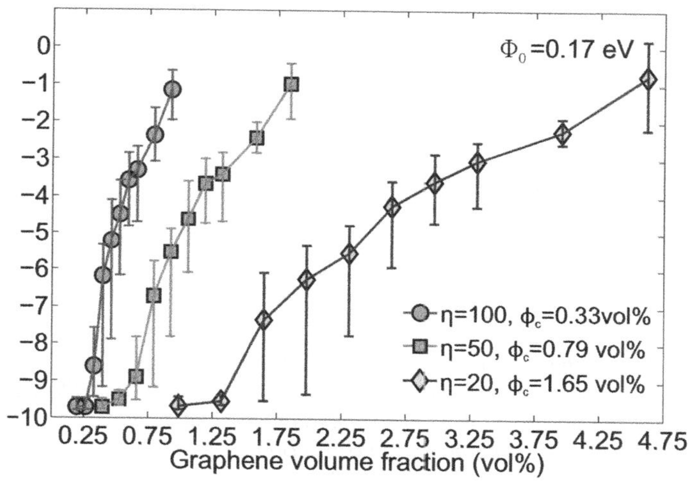

Figure 6.4. *Effective conductivity $\overline{k}_{11}$ as a function of the graphene volume fraction for several graphene aspect ratios η, $\overline{E} = 1.25 \times 10^{-3}$ V/nm, $\Phi_0 = 0.17$ eV. ϕ_c denotes to the percolation threshold. For a color version of this figure, see www.iste.co.uk/bai/nanocomposites.zip*

The side length of the graphene is fixed, and the various aspect ratios are obtained by changing the thickness of the graphene sheets. For each case, the values are averaged over 30 realizations of random distributions of graphene sheets within the RVE. As the behavior of the composite is nonlinear, the results of conductivity are presented for a fixed electric field $\overline{E} = 1.25 \times 10^{-3}$ V/nm. The barrier height is $\Phi_0 = 0.17$ eV. According to our numerical simulations, the effective conductivity clearly depends on the aspect ratio, a larger η provides lower percolation threshold. The obtained percolation thresholds for $\eta = 20, 50,$ and 100 are 1.65 vol%, 0.79 vol%, and 0.33 vol%, respectively.

6.3.3. *Effect of alignment of graphene sheets*

Next, we evaluate the effect of alignment of graphene sheets on the effective conductivity of the composite. For this purpose, we consider, on the one hand, a microstructure with graphene sheets whose positions and orientations are randomly distributed, and, on the other hand, a microstructure where the positions of the graphene sheets are randomly distributed, but the orientation is fixed. Each point corresponds to the mean value over 30 realizations (see realization examples in Figure 6.5).

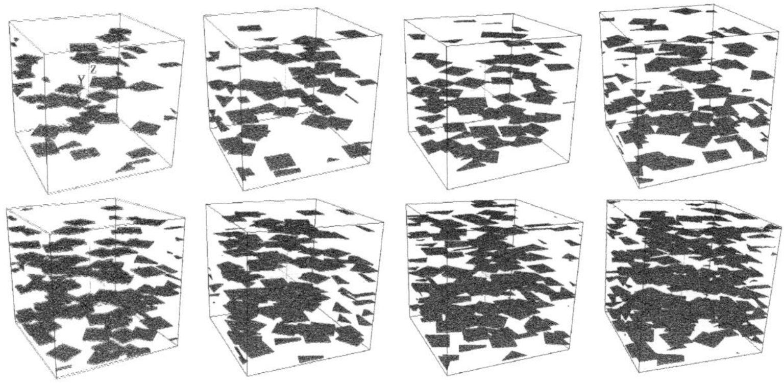

Figure 6.5. *Realizations of microscopic RVE with various graphene volume fraction. The graphene sheets are aligned parallel to the X-Y plane, and the 2D meshes on the graphene sheets are plotted. The microstructure is periodic, a graphene sheet crossing a boundary is extended on the RVE opposite side. a) 0.53 vol%, b) 0.66 vol%, c) 0.79 vol%, d) 0.92 vol%, e) 1.05 vol%, f) 1.19 vol%, g) 1.32 vol% and h) 1.58 vol%. For a color version of this figure, see www.iste.co.uk/bai/nanocomposites.zip*

The results are presented in Figure 6.6. To clearly evidence the anisotropy, we have plotted the three components of the effective conductivity tensor. The parameters are $E = 1.25 \times 10^{-3}$ V/nm, $\Phi_0 = 0.17$ eV and $\eta = 50$.

As expected, the numerical model clearly captures the anisotropic behavior of aligned graphene sheets (see Figure 6.6(b)). Another conclusion is that aligning the graphene sheets does significantly increase either the maximum effective conductivity in the direction, normal to the graphene sheets, or the percolation threshold, as compared to randomly oriented sheets. However, the increase in conductivity after the percolation threshold is sharper in the case of aligned graphene sheets.

Finally, in Figure 6.7 we depict the current density field in the polymer matrix when tunneling effect is taken into account for aligned and randomly oriented graphene sheets to show the percolation path of electric current in both of these configurations. The parameters are $\eta = 50$ and $\Phi_0 = 0.17$ eV, $f = 1.05$ vol%. The electrical field is applied in the X-direction.

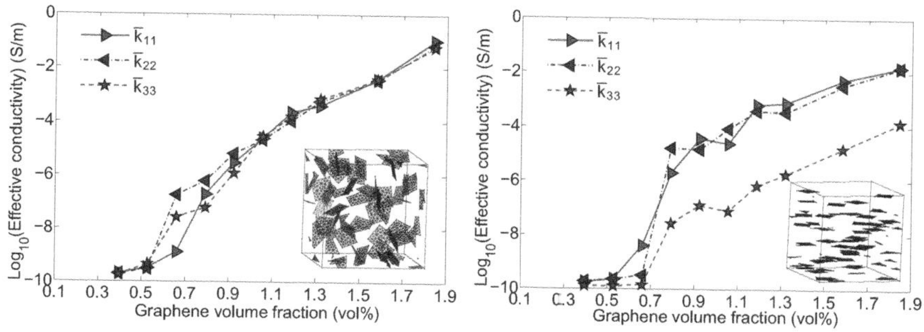

a) b)

Figure 6.6. *Effective conductivity tensor components as a function of the graphene volume fraction: a) random positions and orientations of graphene; b) random positions and direction of graphene normal to the Z-axis. For a color version of this figure, see www.iste.co.uk/bai/nanocomposites.zip*

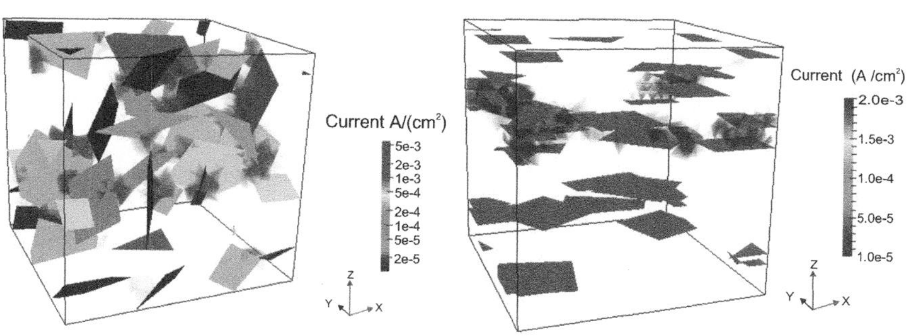

a) b)

Figure 6.7. *Current density in the polymer matrix of microstructure with graphene volume fraction* $f = 1.05$ *vol%: a) random positions and orientations of graphene; b) random positions and direction of graphene normal to the Z-axis. For a color version of this figure, see www.iste.co.uk/bai/nanocomposites.zip*

6.3.4. Comparison between numerical and experimental results

In the following, the obtained numerical results are compared with available experimental data by Stankovich *et al.* (2006) and Zhang *et al.* (2010), respectively. The details of the different material and numerical parameters can be found in Lu *et al.* (2018). The results presented in Figures 6.8 and 6.9 show encouraging agreement between experiments and numerical simulations, even though some discrepancies exist. These disagreements can be explained as follows.

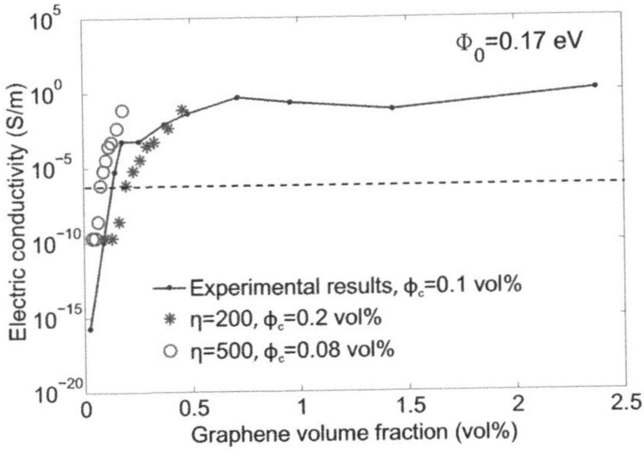

Figure 6.8. *Experimentally measured (Stankovich* et al. *2006) and predicted electric conductivities of polystyrene–graphene composites with various aspect ratio of graphene sheets ($\eta = 200, 500$). ϕ_c denotes the percolation threshold. For a color version of this figure, see www.iste.co.uk/bai/nanocomposites.zip*

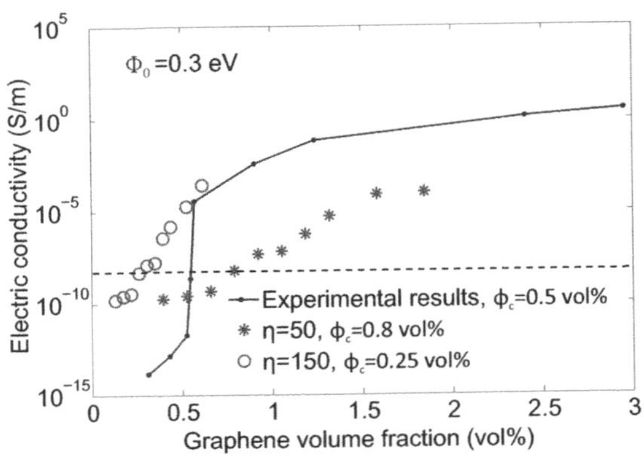

Figure 6.9. *Experimentally measured (Zhang* et al. *2010) and predicted electric conductivities of polyethylene terephthalate/graphene nanocomposites with various aspect ratio of graphene sheets ($\eta = 50, 150$). ϕ_c denotes the percolation threshold. For a color version of this figure, see www.iste.co.uk/bai/nanocomposites.zip*

First, the dispersion state of graphene is hard to control in the experimental preparation, and the aggregation of graphene in the microstructure may lead to the heterogeneity of the macroscopic property. However, in our present work, the numerical simulation is based on randomly distributed graphene sheets. Moreover, the numerical simulation ignores the defects in the nanocomposite, especially the ones on the interface between graphene and the polymer. It should also be noted that we focus on the simulation of nanocomposites with small graphene volume fraction due to the extremely small percolation threshold that we are interested in. The simulation of electric conductivity at high graphene volume fraction calls for a large computation cost and has quite a long computation time due to the refined meshes, so we did not take this further.

6.4. Two-scale approaches

In this section, the former model is used in a structural application. Using a nonlinear model directly obtained from a full RVE simulation in a structure problem, for example, solved by finite elements, is not a trivial problem, as there is no closed form for the effective constitutive law. One direct solution is to numerically evaluate the response of the RVE at each point (Gauss points) in the structural mesh and use a Newton approach to solve the problem (FE^2 method (Feyel 1999)). However, such a procedure is highly time consuming and induces intractable computational times for 3D applications. Here, we present a recent approach (Lu *et al.* 2019), a data-driven-based computational homogenization method based on machine learning, to describe the nonlinear electric conduction in random graphene–polymer nanocomposites. In the proposed technique, the nonlinear effective electric constitutive law is provided by a neural network surrogate model constructed through a learning phase on a set of RVE nonlinear computations. In contrast to multilevel (FE^2) methods where each integration point is associated with a full nonlinear RVE calculation, the nonlinear macroscopic electric field-electric flux relationship is efficiently evaluated by the surrogate neural network model, drastically reducing (by several orders of magnitude) the computational times in multilevel calculations.

Artificial neural networks (ANNs), commonly referred to as *neural networks*, are massively parallel computing systems inspired by the biological neural network of the human brain, that is, an ANN model attempts to use some organizational principles used in a biological neural network, like in the human brain. Thus, ANNs are information-processing mathematical models configured to perform tasks through a learning process and have been extensively used by researchers across different disciplines in order to solve a variety of problems in pattern recognition, prediction, optimization and clustering. One of the most important features of an ANN is its ability to learn the task for which it is designed through examples, provide the network with a set of input/output signals called training samples and use it to create a mapping between the input and the output. Basically, the training of the ANN consists of ordinary steps to tune the synaptic weights so that a map that closely fits the training set is achieved. Three major learning strategies exist: (i) supervised learning, in which each training sample is composed of the input signals and their corresponding outputs, (ii) unsupervised learning, in which prior knowledge of the respective outputs is not required and (iii) reinforced learning, in which the learning algorithms adjust the weights relying on information acquired through the interaction with the model. In this work, we utilize a supervised learning strategy, which is explained in the following. In the present work, the Matlab® Neural Network Toolbox has been used.

6.4.1. *Construction of the surrogate model based on ANN: strategy*

The key idea of this work is to construct a surrogate model based on neural networks in a multilevel finite element framework. In the present method, called neural network computational homogenization (NN-CH), we use a multi-layer perceptron (MLP) to construct the nonlinear relationship between the (input) effective electric field vector $\overline{\mathbf{E}} = [\overline{E}_x, \overline{E}_y, \overline{E}_z]$ and the (output) effective electric current density vector $\overline{\mathbf{J}} = [\overline{J}_x, \overline{J}_y, \overline{J}_z]$ using the model described in section 6.2.2. As this relationship is not known in advance, each set of values for $\overline{\mathbf{E}}$ requires a nonlinear finite element problem to be solved to obtain the corresponding $\overline{\mathbf{J}}$, as defined in section 6.2.2. To avoid these costly calculations at the macro-scale, we propose here to define this relationship through the ANN surrogate model. For this purpose, a set of data consisting of N couples of vectors $\overline{\mathbf{E}}$ and their corresponding $\overline{\mathbf{J}}$ are provided to the neural network.

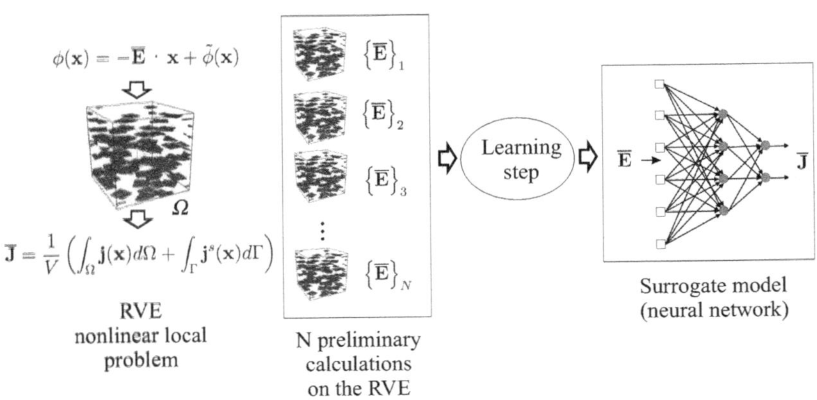

Figure 6.10. *Off-line computational in the data-driven computational homogenization framework. For a color version of this figure, see www.iste.co.uk/bai/nanocomposites.zip*

The procedure for constructing the ANN-based surrogate multiscale model is illustrated in Figures 6.10 and 6.11 and outlined in the following steps:

1) generate N number of electric field vectors $\overline{\mathbf{E}} = [\overline{E}_x, \overline{E}_y, \overline{E}_z]$ to be used for the MLP training;

2) run N FEM nonlinear RVE calculations, for applied effective electric field vectors $\{\overline{\mathbf{E}}\}^{(i)}$, $i = 1, 2, ..., N$ and compute the corresponding effective current densities $\{\overline{\mathbf{J}}\}^{(i)}$, $i = 1, 2, ..., N$, as described in section 6.2.2;

3) train the MLP using $\{\overline{\mathbf{E}}_i\}_{i=1}^N$ as inputs and $\{\overline{\mathbf{J}}_i\}_{i=1}^N$ as the corresponding outputs;

4) check for accuracy of the trained MLP;

5) if the error (using appropriate metrics) is larger than a tolerance, add new input (through steps 1–2) to the set of data and perform steps 3–5 again.

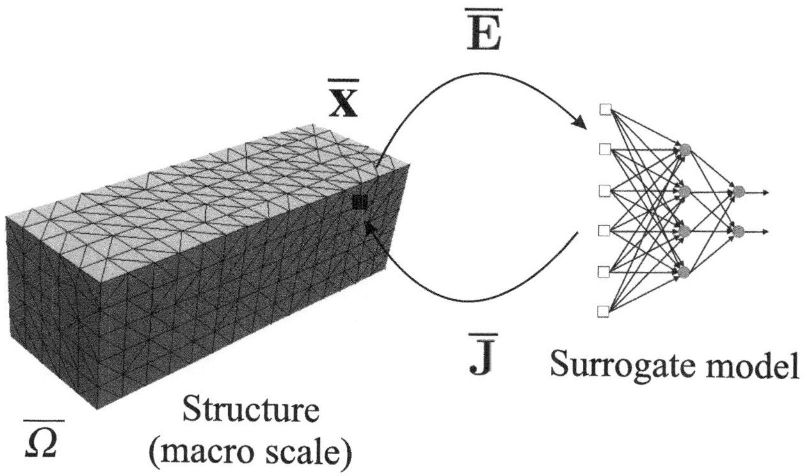

Figure 6.11. *Online computational in the data-driven computational homogenization framework*

6.4.2. *Structural application*

One advantage of the NN-CH approach is that once the surrogate model is constructed, many structural problems can be solved without new calculations on the RVE and taking advantage of the off-line calculation costs. In this next example, we consider the structure described in Figure 6.12. The side lengths of the domain are $L_x = 16$ μm, $L_y = 8$ μm, $L_z = 2$ μm. The diameter of the hole in the middle of the cuboid is $D = 4$ μm.

A mesh of 1820 tetrahedral elements is constructed. The boundary condition is ϕ(x $= 0$ μm) $= 0$ V, ϕ(x$=$ 16 μm) $= 160$ V on the corresponding end surfaces. The current density $\overline{\mathbf{J}} = 0$ on the other boundary surfaces.

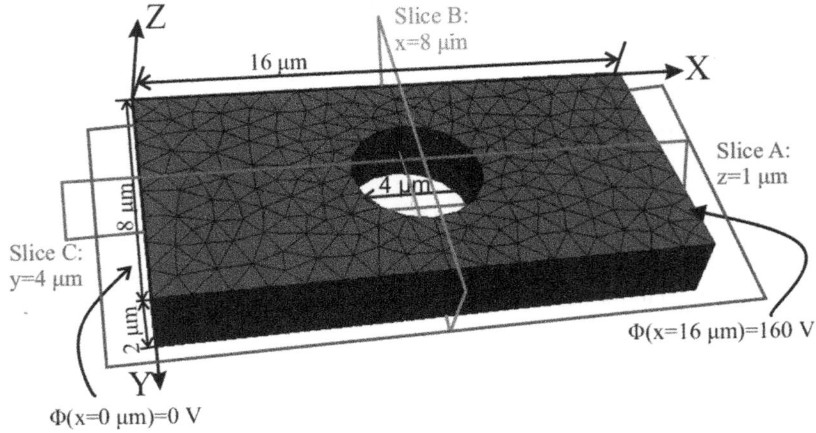

Figure 6.12. *Structural problem: geometry, boundary conditions and mesh. For a color version of this figure, see www.iste.co.uk/bai/nanocomposites.zip*

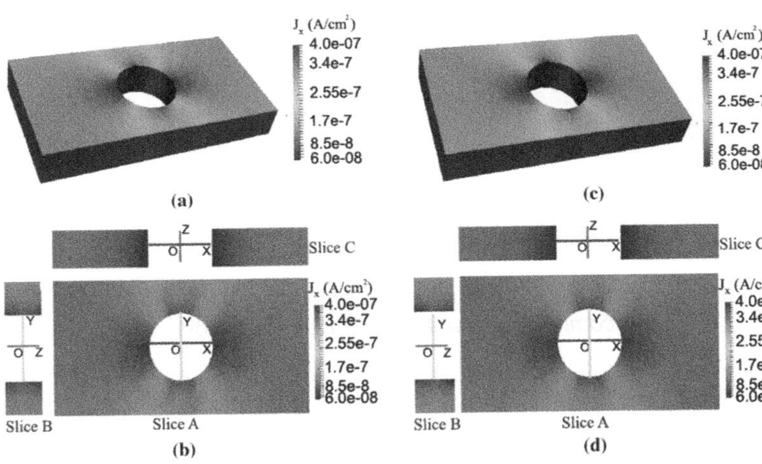

Figure 6.13. *NN-CH simulation results compared with FE² results for the structural composite structure, with* 0.53 *vol% of graphene within the RVE. (a) Overall current density component J_x field and (b) J_x on slices ($x = 8$ μm, $y = 4$ μm, $z = 1$ μm) computed by NN-CH simulation. (c) Overall current density component J_x field, and (d) J_x on slices ($x = 8$ μm, $y = 4$ μm, $z = 1$ μm) computed by FE² simulation. For a color version of this figure, see www.iste.co.uk/bai/nanocomposites.zip*

Results for an RVE of 0.53 vol% are shown in Figure 6.13. For NN-CH, the calculation took 267.45 s with 12 parallel workers, against 126 h for FE^2. A comparison of the different fields by FE^2 is provided in Figure 6.13, showing a very good agreement. Then, we show that accurate nonlinear models based on RVE can be used in structural engineering applications at reasonable costs.

6.5. Electromechanical coupling

In this section, we introduce mechanical effects related to the graphene–polymer interface and its impact on the effective piezoresistive properties of the composite, or in other words, the effect of the mechanical response on the electrical response of the material. For this purpose, a cohesive zone model is identified by atomistic simulations. This cohesive zone model is used to enrich an imperfect interface model, which represents the graphene sheet as an effective imperfect interface. This nonlinear mechanical model is used to generate a deformed representative volume element to study the influence of strain and interfacial decohesion on the conductivity of graphene/polymer nanocomposites. The local model (RVE) employs the nonlinear electrical model described in section 6.2.2.

6.5.1. *Mechanical modeling*

In this section, we use the theory of nonlinear continuum mechanics at finite strains, details of which can be found in previous studies (Gurtin *et al.* (2010), Tadmor *et al.* (2012), among others). We consider the RVE described in Figure 6.2. Here, the RVE in the reference configuration $\Omega_0 \in \mathbb{R}^3$, and the spatial configuration $\Omega_t \in \mathbb{R}^3$.

The two sides of the interface are denoted by Γ_0^+ and Γ_0^-. The unit vector normal to the interface in the reference configuration is $\mathbf{N}(\mathbf{X})$. The displacement of the bulk, and the two sides of the interface are $\mathbf{u}$, $\mathbf{u}^-$ and $\mathbf{u}^+$, respectively. The current positions $\mathbf{x}$ of the material particles at the initial position $\mathbf{X}$ are defined by $\mathbf{x} = \mathbf{X} + \mathbf{u}$ for the bulk. $\mathbf{x}^-$ and $\mathbf{x}^+$ are the current position for the two sides of interface. The graphene sheets are modeled as general imperfect interfaces (Chatzigeorgiou *et al.* 2017, Javili *et al.* 2017), satisfying $[[\mathbf{u}]] = \mathbf{u}^+ - \mathbf{u}^- \neq 0$ and $[[\mathbf{x}]] \neq 0$; the traction through the graphene surfaces is also discontinuous, $[[\mathbf{t}]] = \mathbf{t}^+ - \mathbf{t}^- \neq 0$, where $\mathbf{t}^+$ and $\mathbf{t}^-$ are, respectively, the traction on Γ_0^+ and Γ_0^- associated with the normal $\mathbf{N}(\mathbf{X})$.

Here, we follow the theory of imperfect interface at finite strains, developed by Javili *et al.* (2017). The internal virtual work, δW_{int}, is given in reference configuration by the contributions of polymer bulk and graphene sheets, which are modeled by imperfect interfaces:

$$\delta W_{int}(\mathbf{u}, \delta\mathbf{u}) = \delta W_{int}^{(b)}(\mathbf{u}, \delta\mathbf{u}) + \delta W_{int}^{(s)}(\mathbf{u}, \delta\mathbf{u}) \qquad [6.15]$$

The polymer bulk contribution is

$$\delta W_{int}^{(b)}(\mathbf{u}, \delta\mathbf{u}) = \int_{\Omega_0} \mathbf{P} : \delta\mathbf{F}\, \mathrm{d}V \qquad [6.16]$$

where $\mathbf{P}$ is the first Piola–Kirchoff stress, such as $\mathbf{t} = \mathbf{P} \cdot \mathbf{N}$ and $\delta\mathbf{F} = \nabla_X \delta\mathbf{u}$ is the gradient of the virtual displacement $\delta\mathbf{u}$ with respect to the Lagrangian coordinate $\mathbf{X}$. It should be noted that in this chapter, we defined $(\mathbf{A} \cdot \mathbf{b})_i = A_{ij}b_j$, $\mathbf{a}.\mathbf{b} = a_i b_i$, $(\mathbf{AB})_{ij} = A_{ik}B_{kj}$.

The contribution of imperfect interfaces is given by

$$\delta W_{int}^{(s)}(\mathbf{u}, \delta\mathbf{u}) = \int_{\Gamma_0} \mathbf{P}^{(\mathrm{s})} : \delta\mathbf{F}^{(s)} + \{\{\mathbf{t}\}\}.\, [[\delta\mathbf{u}]]\, \mathrm{d}S \qquad [6.17]$$

where $\mathbf{P}^{(s)}$ is surface first Piola–Kirchof stress and $\delta\mathbf{F}^{(s)} = \nabla_X \delta\mathbf{u}^{\mathrm{s}} \cdot \mathbf{I}_0^{\mathrm{s}}$ denotes the surface gradient of the virtual displacement, where $\delta\mathbf{u}^{\mathrm{s}} = 1/2\,(\delta\mathbf{u}^+ + \delta\mathbf{u}^-)$ is a virtual displacement field on the interface and $\mathbf{I}_0^{\mathrm{s}} = \mathbf{I} - \mathbf{N} \otimes \mathbf{N}$ is the projector into the tangent plan of the interface in the reference configuration. $\{\{\mathbf{t}\}\} = 1/2\,(\mathbf{t}^+ + \mathbf{t}^-)$ is the average of traction. Note that the expression of internal virtual work with imperfect interfaces is true by constraining the interface motion to the midplane (Javili *et al.* 2017), that is, by imposing the definition of the interface displacement $\mathbf{u}^{\mathrm{s}} = 1/2\,(\mathbf{u}^+ + \mathbf{u}^-)$.

Using the second Piola–Kirchhoff stress tensor $\mathbf{S} = \mathbf{F}^{-1}\mathbf{P}$ (respectively, the surface second Piola–Kirchhoff stress tensor, $\mathbf{S}^s = (\mathbf{F}^s)^{-1}\mathbf{P}^s$ with $\mathbf{F}^s = \mathbf{I}_0^{\mathrm{s}} + \nabla_X \mathbf{u}^{\mathrm{s}} \cdot \mathbf{I}_0^{\mathrm{s}}$), we finally obtain the following expression of internal virtual work (for a configuration in static equilibrium):

$$\delta W_{int}^{(b)}(\mathbf{u}, \delta\mathbf{u}) = \int_{\Omega_0} \mathbf{S} : \delta\boldsymbol{\varepsilon}\, \mathrm{d}V \qquad [6.18]$$

$$\delta W_{int}^{(s)}(\mathbf{u}, \delta\mathbf{u}) = \int_{\Gamma_0} \mathbf{S}^{\mathrm{s}} : \delta\boldsymbol{\varepsilon}^{s} + \{\{\mathbf{t}\}\}.\, [[\delta\mathbf{u}]]\, \mathrm{d}S \qquad [6.19]$$

where $\delta\boldsymbol{\varepsilon}$ (respectively, $\delta\boldsymbol{\varepsilon}^{s}$) is the variation of the symmetric Green–Lagrange strain tensor, $\boldsymbol{\varepsilon} = \frac{1}{2}(\mathbf{F}^T\mathbf{F} - \mathbf{I})$ (respectively, the variation of the surface Green–Lagrange strain tensor, $\boldsymbol{\varepsilon}^{s} = \frac{1}{2}((\mathbf{F}^{\mathrm{s}})^T\,\mathbf{F}^{\mathrm{s}} - \mathbf{I}_0^{\mathrm{s}})$).

6.5.2. *Constitutive laws*

We choose the isotropic Saint Venant–Kirchhoff model for the bulk part, which is an extension of the linear elastic material model. The second Piola–Kirchoff stress is given by:

$$\mathbf{S} = 2\mu^{(b)}\varepsilon + \lambda^{(b)}\left(\varepsilon : \mathbf{I}\right)\mathbf{I} \qquad [6.20]$$

where the Lame's coefficients are $\lambda^{(b)} = 6890$ MPa and $\mu^{(b)} = 680$ MPa (Lu 2017). They are identified by deformation of MD simulation box of pure PE following classical procedures (Theodorou and Suter 1986).

As for the bulk, for sake of simplicity, we assume that the behavior of the imperfect interface is reversible. The surface elastic behavior of graphene is assumed to be isotropic along its tangent plane to the surface, such that the surface second Piola–Kirchoff stress is also given by the Saint Venant–Kirchhoff model:

$$\mathbf{S}^{\mathrm{s}} = 2\mu^{(s)}\varepsilon^{s} + \lambda^{(s)}\left(\varepsilon^{s} : \mathbf{I}_0^{\mathrm{s}}\right)\mathbf{I}_0^{\mathrm{s}} \qquad [6.21]$$

where the surface Lame's coefficient $\lambda^{(s)} = 19.0$ N.m^{-1} and $\mu^{(s)} = 18.7$ N.m^{-1} are identified by the MD simulation (Lu 2017).

The expression of traction $\{\{\mathbf{t}\}\}$ is assumed to be aligned with the displacement jump $[[\mathbf{u}]]$ and is given by:

$$\{\{\mathbf{t}\}\} = g_{cz}\left(||[[\mathbf{u}]]||\right)\frac{[[\mathbf{u}]]}{||[[\mathbf{u}]]||} \qquad [6.22]$$

where $g_{cz}(\cdot)$ is the function identified by the MD simulations in equation [6.27].

6.5.3. *Weak form of mechanical problem*

In order to relate the micro stress and strain fields to the imposed macroscopic strain, the effective quantities are defined as in Javili *et al.* (2017):

$$\overline{\mathbf{F}} = \frac{1}{||\Omega_0||}\int_{\Omega_0}\mathbf{F}\,\mathrm{d}V + \frac{1}{||\Omega_0||}\int_{\Gamma_0}[\![\mathbf{u}]\!]\otimes\mathbf{N}(\mathbf{X})\,\mathrm{d}S \qquad [6.23]$$

and

$$\overline{\mathbf{P}} = \frac{1}{||\Omega_0||}\int_{\Omega_0}\mathbf{P}\,\mathrm{d}V + \frac{1}{||\Omega_0||}\int_{\Gamma_0}\mathbf{P}^{\mathrm{s}}\,\mathrm{d}S \qquad [6.24]$$

where $\overline{\mathbf{F}}$ and $\overline{\mathbf{P}}$ are the effective deformation gradient and effective first Piola–Kirchoff stress, respectively. The effective Green–Lagrange strain tensor $\overline{\varepsilon}$ and the effective second Piola–Kirchoff stress $\overline{\mathbf{S}}$ are defined by:

$$\overline{\varepsilon} = \frac{1}{2}\left(\overline{\mathbf{F}}^T\overline{\mathbf{F}} - \mathbf{I}\right) \quad \text{and} \quad \overline{\mathbf{S}} = \overline{\mathbf{F}}^{-1}\overline{\mathbf{P}} \qquad [6.25]$$

To impose that the incremental internal virtual work at the micro-scale, δW_{int}, is equal to the macroscopic internal virtual work, $\overline{\delta W}_{int} = \overline{\mathbf{P}} : \delta\overline{\mathbf{F}}$, we chose periodic boundary conditions to satisfy the extended Hill–Mandel condition (Javili *et al.* 2017). The weak form associated with the mechanical problem at micro-scale is given by: find $\mathbf{u} \in H^1(\Omega_0)$, satisfying the boundary conditions $\mathbf{u} = \overline{\mathbf{F}}.\mathbf{x} + \tilde{\mathbf{u}}$ over $\partial\Omega_0$ with $\tilde{\mathbf{u}}$ periodic, such that

$$\int_{\Omega_0} \mathbf{S}(\mathbf{u}) : \delta\varepsilon\left(\mathbf{u}, \delta\mathbf{u}\right) \, \mathrm{d}V$$

$$+ \int_{\Gamma_0} \mathbf{S}^s(\mathbf{u}) : \delta\varepsilon^s\left(\mathbf{u}, \delta\mathbf{u}\right) + \{\{\mathbf{t}([[\mathbf{u}]])\}\}.[[\delta\mathbf{u}]] \, \mathrm{d}S = 0 \qquad [6.26]$$

for all $\delta\mathbf{u} \in H_0^1(\Omega_0)$.

We use FEM to discretize the solution space, linear tetrahedrons for the bulk part and linear triangles for interfaces. A Newton–Raphson procedure is used to solve this nonlinear problem step by step for small increments $\Delta\overline{\mathbf{F}}$ of effective strain gradient.

6.5.4. *Identification of cohesive zone model*

To study the separation in opening mode, graphene is moved in successive steps of 0.5Å along the Z direction by following a minimization procedure. The graphene atoms and bottom layer of the polymer are kept fixed (see Figure 6.14(a)). The separation process is depicted in Figure 6.14. The polymer chains undergo stretch at the beginning along the Z direction, then form highly oriented structures, called fibrils or nano-fibrils. Voids appear between the fibrils during the decohesion. This deformation mechanism observed during the simulation is similar to the nano-crazes of some semi-crystalline polymers, such as polybutene (Thomas *et al.* 2009, Detrez 2008). The size of the void grows along the separation direction and the extended chains slide along the graphene sheet to increase the fibrils, as described in the reference review paper by Kramer and Berger (1990). It should be noted that the separation is controlled by the chain desorption at the graphene surface by sliding, which is dominated by Van der Waals interactions.

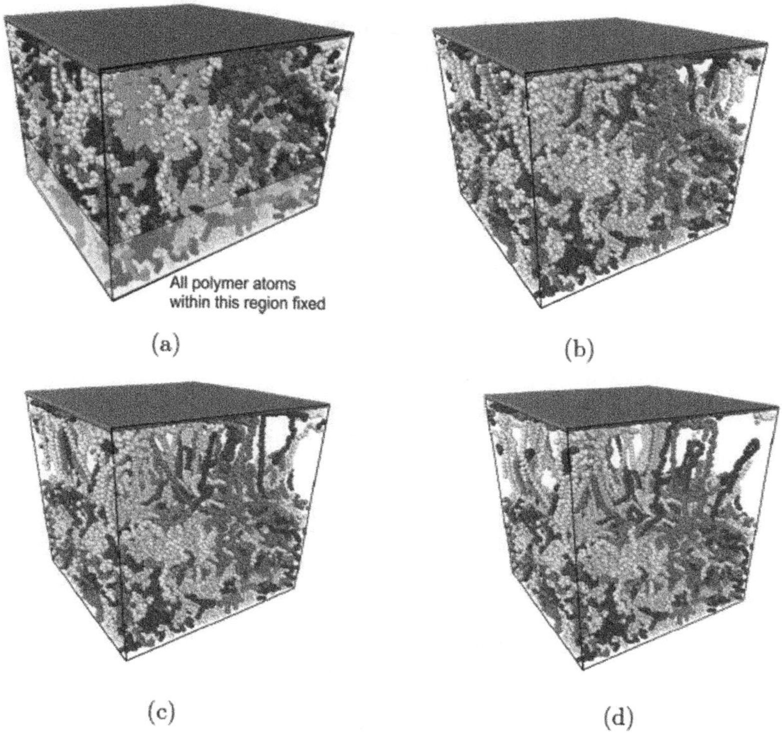

Figure 6.14. *Evolution of the atomistic model during the normal separation. The model contains 44,860 atoms, and the graphene is moved with a step of 0.5 Å. The graphene atoms and the bottom layer of the polymer are fixed during the relaxation. For a color version of this figure, see www.iste.co.uk/bai/nanocomposites.zip*

The average force of polymer on graphene was monitored, from which we can get the normal traction force t_n of the cohesive zone as a function of the displacement of the graphene layer, $[[u_n]]$, as shown in Figure 6.15. First, the force varies linearly with the displacement of the graphene sheet. Then, the curve bends to reach a maximum at 0.7 nm, called the yield threshold. This phase corresponds to the nano-fribils creation and to the cavity initiation. Once the yield threshold is crossed, the force decreases with the displacement of the graphene sheet. During this phase, the chains slip on the graphene sheet to feed the fibrils. It is interesting to note that the observed softening is related to the reduction of the contact area between the polymer chains and graphene. We propose fitting the MD results with the following empirical model:

$$t_n = g_{cz}\left([[u_n]]\right) = \begin{cases} -1529[[u_n]]^2 + 2150[[u_n]] & \text{if } 0 \leq [[u_n]] < 0.7 \\ \frac{65}{[[u_n]]^8} - \frac{4.31}{[[u_n]]^{14}} + 263.74 & \text{if } 0.7 \leq [[u_n]] \leq 1.15 \\ 360\exp(-0.16[[u_n]]) - 15.12 & \text{if } [[u_n]] > 1.15 \end{cases} \quad [6.27]$$

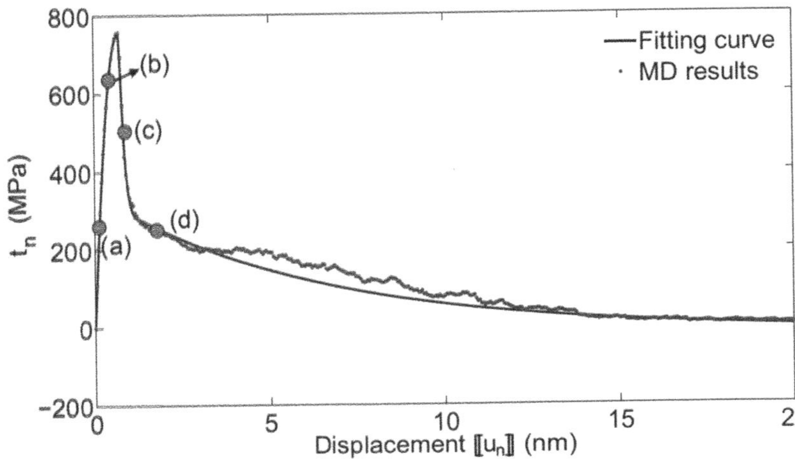

Figure 6.15. *Traction force, t_n versus displacement of graphene layer $[[u_n]]$. The points denote the MD results and the line is the fitting curve. There is a correspondence between the bold points with the numbers (1–4) on the curve and the figures above. For a color version of this figure, see www.iste.co.uk/bai/nanocomposites.zip*

The above model is incorporated in the mechanical model described in section 6.5.1 and is coupled with the nonlinear electrical model described in section 6.2.2.

6.5.5. *Evolution of electrical properties under stretching of the composite*

Using the above electro-mechanical model of nanocomposites, we study the impact of both stretching and decohesion at the graphene polymer interface on the electric conductivity. Indeed, the conductivity is controlled by the tunneling effect, which depends strongly on the distance between graphene. Therefore, we impose an macroscopic elongation, $\overline{\varepsilon} = \overline{\varepsilon}_{11}\mathbf{e}_1 \otimes \mathbf{e}_1$, on the RVEs, from $\overline{\varepsilon}_{11} = 0\%$ to 10%. Due to the large computational times, only one RVE microstructure is randomly studied for each graphene volume fraction.

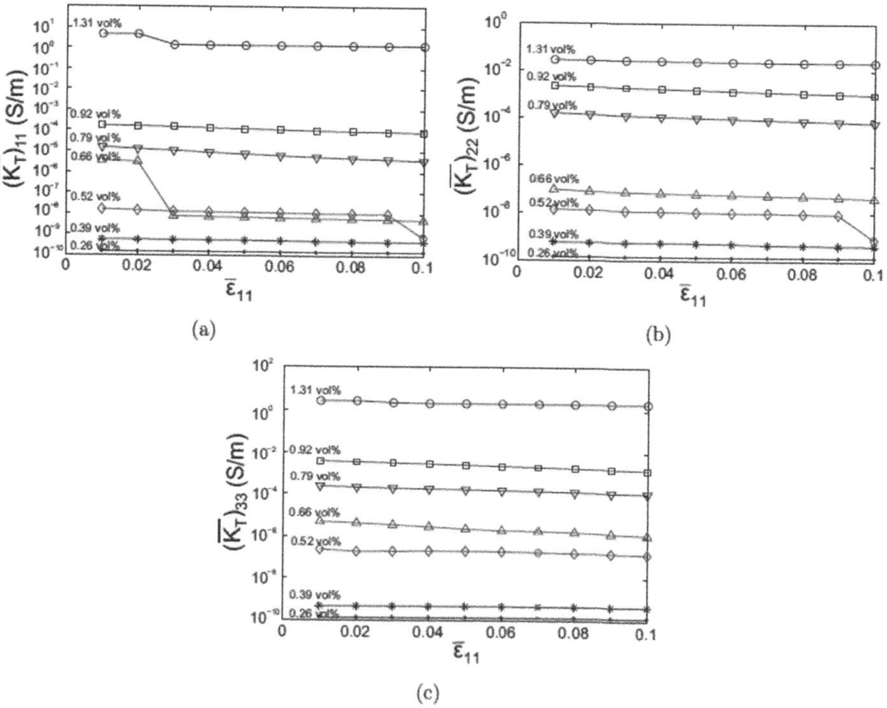

Figure 6.16. *Effective electrical conductivity of graphene reinforced nanocomposites as a function of the deformation for various graphene volume fraction. The barrier height between graphene and polymer matrix is set to 0.17 eV, and the graphene aspect ratio is 75. The applied electric field is 0.0025 V/nm. a) $(\overline{K}_T)_{11}$, b) $(\overline{K}_T)_{22}$ and c) $(\overline{K}_T)_{33}$. For a color version of this figure, see www.iste.co.uk/bai/nanocomposites.zip*

The deformed microstructures are stored for each 1% increment of deformation and the distance function, $d(\mathbf{x})$, is updated. Introducing the new distance function, the electrical conductivities at different effective strain $\overline{\varepsilon}_{11}$ are shown in Figure 6.16. The boundary between the insulator and conductor is defined as 10^{-8} S/m (Singh 2011), below which the material is supposed to be the insulator. In the other situation, it behaves like a conductor. Focusing on the electrical conductivity along the direction of deformation $(\overline{K}_T)_{11}$, in Figure 6.16(a) we can observe that the mechanical deformation has little effect on the electrical conductivity of the nanocomposites when the graphene volume fraction is below the percolation threshold, $f < f_c = 0.52$ vol%. When the graphene volume fraction is above the percolation threshold ($f > f_c = 0.52$ vol%), the electrical conductivity $(\overline{K}_T)_{11}$ decreases with the applied elongation, but it should be noted that the nanocomposites remain as the conductor. However, if the graphene volume fraction is around the percolation threshold ($f \approx f_c = 0.52$ vol%), a sharp decrease in the electrical conductivity can be seen when the nanocomposites are subjected to strain, which is regarded as a transition point from conductor to insulator. For instance, with 0.66 vol% graphene the transition point of the sample is $\bar{\varepsilon}_{11} \approx 3\%$, and with $f = f_c = 0.52$ vol% graphene it is $\bar{\varepsilon}_{11} \approx 10\%$. We note that this drop in conductivity does not occur for any curve above a volume fraction $f \geq 0.79$ vol%. However, the elongation would not affect the electrical conductivities of the nanocomposites in the transverse directions $(\overline{K}_T)_{22}$ and $(\overline{K}_T)_{33}$, which can be seen from Figures 6.16(b) and (c). This is probably a consequence of the fact that the simulations for each volume fraction have only been performed on one realization, and that some configurations might be more favorable to this effect. The results demonstrate that the graphene nanocomposite with certain graphene volume fraction (around percolation threshold) may lead to the existence of conductor to insulator behavior under finite strain. But it is also dependent on the microstructure of nanocomposite, and it only happens to those whose current path formed by graphene can be destroyed during the tension.

After observing this typical conductor-to-insulator transition for the composite with $f = 0.66$ vol% graphene by the proposed model, we compare the effective conductivity with the results that are estimated without considering the decohesion between graphene and the polymer matrix (i.e. we prescribe $[[\mathbf{u}]] = 0$). It can be seen in Figure 6.17 that by neglecting the cohesive interface, the transition point increases from $\bar{\varepsilon}_{11} \approx 3\%$ to 8%, which shows the important role of decohesion at the interface in predicting the piezoresistivity properties of polymer graphene nanocomposites. It is interesting to note that it is theoretically possible to design a composite that can go from conductor to insulator by varying the applied strain on the system. This transition can be induced mainly by the decohesion for weak interfaces, or only by strain for a stronger interface, but for a more important applied elongation.

6.6. Conclusion

In this chapter, we have presented several recent advances in the modeling of graphene–polymer nanocomposites. First, a model at the scale of an RVE was presented, where the effects of microstructures and nonlinearities associated with the quantum tunneling effects have been taken into account. This model can be used to capture the effective nonlinear electrical properties of the composites with respect to the microstructural parameters (graphene platelets orientation, volume fraction, etc.). Then, a two-scale numerical strategy was introduced to use this model at the scale of a real structure through a numerical homogenization process involving machine learning. Finally, we have presented the effects of weak electromechanical coupling in these composites by introducing the mechanical damage behavior at the interfaces, which have been identified by molecular dynamics. These different numerical tools provide promising modeling approaches to predict and design the electro-mechanical behavior of future nanocomposites.

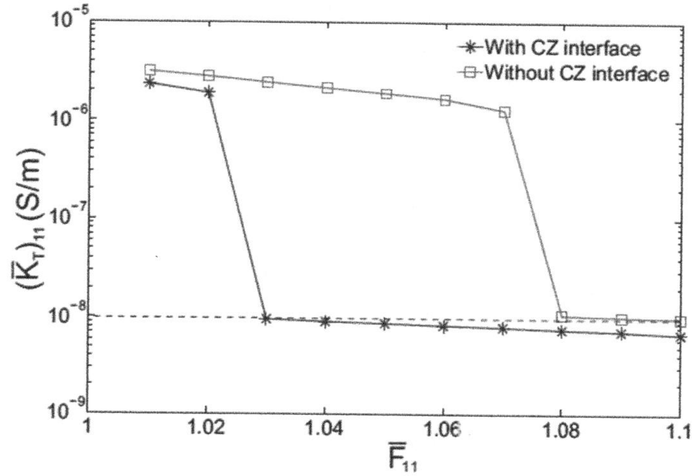

Figure 6.17. *Effective electric conductivity* $(\overline{K_T})_{11}$ *as a function of effective strain* $\bar{\epsilon}_{11}$ *for the composite with* 0.66 *vol%* *graphene, both with and without considering the cohesive interface,* $\Phi_0 = 0.17$. *For a color version of this figure, see www.iste.co.uk/bai/nanocomposites.zip*

6.7. References

Ambrosetti, G., Grimaldi, C., Balberg, I., Maeder, T., Danani, A., Ryser, P. (2010). Solution of the tunneling-percolation problem in the nanocomposite regime. *Physical Review B*, 81(15), 155434.

Ansari, S. and Giannelis, E.P. (2009). Functionalized graphene sheet-poly (vinylidene fluoride) conductive nanocomposites. *Journal of Polymer Science Part B: Polymer Physics*, 47(9), 888–897.

Bauhofer, W. and Kovacs, J.Z. (2009). A review and analysis of electrical percolation in carbon nanotube polymer composites. *Composites Science and Technology*, 69(10), 1486–1498.

Britnell, L., Ribeiro, R.M., Eckmann, A., Jalil, R., Belle, B.D., Mishchenko, A., Kim, Y., Gorbachev, R.V., Georgiou, T., Morozov, S.V., Grigorenko, A.N., Geim, A.K., Casiraghi, C., CastroNeto, A.H., Novoselov, K.S. (2013). Strong light-matter interactions in heterostructures of atomically thin films. *Science*, 340(6138), 1311–1314.

Bryning, M.B., Islam, M.F., Kikkawa, J.M., Yodh, A.G. (2005). Very low conductivity threshold in bulk isotropic single-walled carbon nanotube–epoxy composites. *Adv. Mater.*, 17(9), 1186–1191.

Castaneda, P. and Willis, J. (1995). The effect of spatial distribution on the effective behavior of composite materials and cracked media. *Journal of the Mechanics and Physics of Solids*, 43(12), 1919–1951.

Charlaix, E. (1986). Percolation threshold of a random array of discs: A numerical simulation. *Journal of Physics A: Mathematical and General*, 19(9), L533.

Chatzigeorgiou, G., Meraghni, F., Javili, A. (2017). Generalized interfacial energy and size effects in composites. *J. Mech. Phys. Solids*, 106, 257–282.

Chiu, F.C. (2014). A review on conduction mechanisms in dielectric films. *Advances in Materials Science & Engineering*, 578168, 1–18.

Detrez, F. (2008). Nanomécanismes de déformation des polymères semi-cristallins : étude in situ par microscopie à force atomique et modélisation. PhD Thesis, Université Lille I.

Du, F., Fisher, J., Winey, K. (2005). Effect of nanotube alignment on percolation conductivity in carbon nanotube/polymer composites. *Physical Review B*, 72(12), 121404.

El-Kady, M.F. and Kaner, R.B. (2013). Scalable fabrication of high-power graphene micro-supercapacitors for flexible and on-chip energy storage. *Nature Communications*, 4, 1475.

Fan, P., Wang, L., Yang, J., Chen, F., Zhong, M. (2012). Graphene/poly(vinylidene fluoride) composites with high dielectric constant and low percolation threshold. *Nanotechnology*, 23(36), 365702.

Fan, Z., Gong, F., Nguyen, S.T., Duong, H.M. (2015). Advanced multifunctional graphene aerogel–poly (methyl methacrylate) composites: Experiments and modeling. *Carbon*, 81, 396–404.

Feyel, F. (1999). Multiscale FE^2 elastoviscoplastic analysis of composite structure. *Computational Material Science*, 16(1–4), 433–454.

Gao, C., Zhang, S., Wang, F., Wen, B., Han, C., Ding, Y., Yang, M. (2014). Graphene networks with low percolation threshold in ABS nanocomposites: Selective localization and electrical and rheological properties. *ACS Applied Materials & Interfaces*, 6(15), 12252–12260.

Grimaldi, C. and Balberg, I. (2006). Tunneling and nonuniversality in continuum percolation systems. *Phys. Rev. Lett.*, 96(6), 066602.

Guo, W., Cheng, C., Wu, Y., Jiang, Y., Gao, J., Li, D., Jiang, L. (2013). Bio-inspired two-dimensional nanofluidic generators based on a layered graphene hydrogel membrane. *Advanced Materials*, 25(42), 6064–6068.

Gurtin, M.E., Fried, E., Anand, L. (2010). *The Mechanics and Thermodynamics of Continua*. Cambridge, New York.

Haggenmueller, R., Guthy, C., Lukes, J.R., Fischer, J.E., Winey, K.I. (2007). Single wall carbon nanotube/polyethylene nanocomposites: Thermal and electrical conductivity. *Macromolecules*, 40(7), 2417–2421.

Hamann, C., Burghardt, H., Frauenheim, T. (1988). *Electrical Conduction Mechanisms in Solids*. VEB Deutscher Verlag der Wissenschaften, Berlin.

He, L. and Tjiong, S.C. (2013). Low percolation threshold of graphene/polymer composites prepared by solvothermal reduction of graphene oxide in the polymer solution. *Nanoscale Research Letters*, 8(1), 132.

Hicks, J., Behnamand, A., Ural, A. (2009). A computational study of tunneling-percolation electrical transport in graphene-based nanocomposites. *Applied Physics Letters*, 95(21), 213103.

Javili, A., Steinmann, P., Mosler, J. (2017). Micro-to-macro transition accounting for general imperfect interfaces. *Comput. Methods Appl. Mech. Engrg.*, 317, 274–317.

Jiang, J.Y., Kim, M.S., Jeong, H.M., Shon, C.M. (2009). Graphite oxide/poly (methyl methacrylate) nanocomposites prepared by a novel method utilizing macroazoinitiator. *Composites Science and Technology*, 69(2), 186–191.

Kim, H. and Macosko, C.W. (2009). Processing-property relationships of polycarbonate/graphene composites. *Polymer*, 50(15), 3797–3809.

Kim, K.S., Zhao, Y., Jang, H., Lee, S.Y., Kim, J.M., Kim, K.S., Ahn, J.H., Kim, P., Choi, J.Y., Hong, B.H. (2009). Large-scale pattern growth of graphene films for stretchable transparent electrodes. *Nature*, 457(7230), 706.

Kramer, E.J. and Berger, L.L. (1990). Fundamental processes of craze growth and fracture. *Polym. Sci.*, 91–92, 1–68.

Krishnan, S., Stefanakos, E., Bhansali, S. (2008). Effects of dielectric thickness and contact area on current-voltage characteristics of thin film metal-insulator-metal diodes. *Thin Solid Films*, 516(8), 2244–2250.

Liang, J., Wang, Y., Huang, Y., Ma, Y., Liu, Z., Cai, J., Zhang, C., Gao, H., Chen, Y. (2009). Electromagnetic interference shielding of graphene/epoxy composites. *Carbon*, 47(3), 922–925.

Lu, X. (2017). Multiscale electro-mechanical modeling of graphene/polymer nanocomposites. PhD Thesis, Université Paris-Saclay.

Lu, X., Yvonnet, J., Detrez, F., Bai, J. (2018). Low electrical percolation thresholds and nonlinear effects in graphene-reinforced nanocomposites: A numerical analysis. *Journal of Composite Materials*, 52(20), 2767–2775.

Lu, X., Giovanis, D., Yvonnet, J., Papadopoulos, V., Detrez, F., Bai, J. (2019) A data-driven computational homogenization method based on neural networks for the nonlinear anisotropic electrical response of graphene/polymer nanocomposites. *Computational Mechanics*, 64(2), 307–321.

Maity, N.P., Maity, R., Thapa, R.K., Baishya, S. (2016). A tunneling current density model for ultra thin HfO2 high-k dielectric material based {MOS} devices. *Superlattices and Microstructures*, 95, 24 – 32.

Mannoor, M.S., Tao, H., Clayton, J.D., Sengupta, A., Kaplan, D.L., Naik, R.R., Verma, N., Omenetto, F.G., McAlpine, M.C. (2012). Graphene-based wireless bacteria detection on tooth enamel. *Nature Communications*, 3, 763.

Novoselov, K.S., Geim, A.K., Morozov, S.V., Jiang, D., Zhang, Y., Dubonos, S.V., Grigorieva, I.V., Firsov, A.A. (2004). Electric field effect in atomically thin carbon films. *Science*, 306(5696), 666–669.

Oskouyi, A.B., Sundararaj, U., Mertiny, P. (2014). Current-voltage characteristics of nanoplatelet-based conductive nanocomposites. *Nanoscale Research Letters*, 9(1), 369.

Otten, R.H. and van der Schoot, P. (2009). Continuum percolation of polydisperse nanofillers. *Phys. Rev. Lett.*, 103(22), 225704.

Otten, R.H. and van der Schoot, P. (2011). Connectivity percolation of polydisperse anisotropic nanofillers. *The Journal of Chemical Physics*, 134(9), 094902.

Qi, X., Yan, D., Jiang, Z., Cao, Y., Yu, Z., Yavari, F., Koratkar, N. (2011). Enhanced electrical conductivity in polystyrene nanocomposites at ultra-low graphene content. *ACS Applied Materials & Interfaces*, 3(8), 3130–3133.

Sheng, P. (1980). Fluctuation-induced tunneling conduction in disordered materials. *Physical Review B*, 21(6), 2180.

Simmons, J.G. (1963). Electric tunnel effect between dissimilar electrodes separated by a thin insulating film. *Journal of Applied Physics*, 34(9), 2581–2590.

Simmons, J.G. (1971). Conduction in thin dielectric films. *Journal of Physics D: Applied Physics*, 4(5), 613.

Singh, A.K. (2011). *Electronic Devices and Integrated Circuits*. PHI Learning Pvt. Lt, New Delhi.

Stankovich, S., Dikin, D.A., Dommett, G.H., Kohlhaas, K.M., Zimney, E.J., Stach, E.A., Piner, R.D., Nguyen, S.T., Ruoff, R.S. (2006). Graphene-based composite materials. *Nature*, 442(7100), 282–286.

Tadmor, E.B., Miller, R.E., Elliott, R.S. (2012). *Continuum Mechanics and Thermodynamics from Fundamental Concepts to Governing Equations*. Cambridge University Press, New York.

Theodorou, D.N. and Suter, U.W. (1986). Atomistic modeling of mechanical properties of polymeric glasses. *Macromolecules*, 19(1), 139–154.

Thomas, C., Seguela, R., Detrez, F., Miri, V., Vanmansart, C. (2009). Plastic deformation of spherulitic semi-crystalline polymers: An *in situ* AFM study of polybutene under tensile drawing. *Polymer*, 50(15), 3714–3723.

Tkalya, E., Ghislandi, M., Otten, R., Lotya, M., Alekseev, A., van der Schoot, P., Coleman, J., de With, G., Koning, C. (2014). Experimental and theoretical study of the influence of the state of dispersion of graphene on the percolation threshold of conductive graphene/polystyrene nanocomposites. *ACS Applied Mater. Interfaces*, 6(17), 15113–15121.

Trionfi, A., Wang, D., Jacobs, J., Tan, L., Vaia, R., Hsu, J. (2009). Direct measurement of the percolation probability in carbon nanofiber-polyimide nanocomposites. *Phys. Rev. Lett.*, 102(11), 116601.

Wang, Y., Weng, G.J., Meguid, S.A., Hamouda, A.M. (2014). A continuum model with a percolation threshold and tunneling-assisted interfacial conductivity for carbon nanotube-based nanocomposites. *Journal of Applied Physics*, 115(19), 193706.

Wang, Y., Shan, J.W., Weng, G.J. (2015). Percolation threshold and electrical conductivity of graphene-based nanocomposites with filler agglomeration and interfacial tunneling. *Journal of Applied Physics*, 118(6), 065101.

Xia, X., Wang, Y., Zhong, Z., Weng, G.J. (2017). A frequency-dependent theory of electrical conductivity and dielectric permittivity for graphene-polymer nanocomposites. *Carbon*, 111, 221–230.

Yousefi, N., Gudarzi, M.M., Zheng, Q., Aboutalebi, S.H., Sharif, F., Kim, J.K. (2012). Self-alignment and high electrical conductivity of ultralarge graphene oxide–polyurethane nanocomposites. *Journal of Materials Chemistry*, 22(25), 12709–12717.

Yvonnet, J., He, Q.-C., Toulemonde, C. (2008). Numerical modelling of the effective conductivities of composites with arbitrarily shaped inclusions and highly conducting interface. *Composites Science and Technology*, 68(13), 2818–2825.

Zhang, H., Zheng, W., Yan, Q., Yang, Y., Wang, J., Lu, Z., Ji, G., Yu, Z. (2010). Electrically conductive polyethylene terephthalate/graphene nanocomposites prepared by melt compounding. *Polymer*, 51(5), 1191–1196.

7

Computational Modeling of Carbon Nanofiller Networks in Polymer Composites

Angel MORA
CICY, Mérida, Mexico

7.1. Introduction

Conductive nanofillers are typically added into polymers to improve their electrical properties. Carbon-based nanofillers such as carbon nanotubes (CNTs), graphene nanoplatelets (GNPs) or carbon black (CB) are commonly used as conductive nanofillers (Zhou and Lubineau 2013, Li and Zhong 2011, Ezquerra *et al.* 1986). The resulting nanocomposite material can be used in a wide range of applications. For instance, the addition of GNPs into a polymer can reduce its dielectric properties, which reduces the accumulation of electrical charge (Ge and Cosgrove 2015). This prevents electronic equipment from suffering damage due to an electric discharge. Polymers filled with conductive nanofillers possess piezoresistive properties that may be exploited for strain sensing and structural health monitoring applications (Li *et al.* 2008b, Nanni *et al.* 2012, Wernik and Meguid 2010). Table 7.1 shows some examples of the wide range of applications and the type of conductive material obtained depending on the electrical conductivity of polymer nanocomposites.

The change in behavior of a polymer from insulator to electrical conductor by the addition of conductive nanofillers is explained by the percolation theory (Stauffer and Aharony 1994). Electrical percolation is illustrated in the schematic in Figure 7.1,

Nanocomposites,
coordinated by Jinbo BAI. © ISTE Ltd 2022.

where the composite's conductivity is shown as a function of nanofiller content. The insets of that figure show schematics of nanofiller networks at different volume contents, where CNTs are used as an example. When the nanofiller content is low, there is a slight increase in electrical conductivity. Then, when a critical concentration is reached, known as the percolation threshold, a sharp increase in conductivity is observed. The percolation threshold is denoted as ϕ_C in Figure 7.1. At a nanofiller content equal to the percolation threshold, a network of connected nanofillers that spans the whole composite is formed. This network is known as the percolated network and is shown in light gray in the middle and right insets in Figure 7.1. It is the percolated network that makes the conduction of electricity possible, as electrons mainly move along nanofillers and from one nanofiller to the other. Thus, the electrical conductivity of the composite is largely defined by the conductivity of the percolated network.

Conductivity [S/cm]	Classification	Application
10^4–10^{-1}	Highly conductive	Conductors (e.g. metal replacements, resistors, thermoelectric materials)
10^{-1}–10^{-6}	Conductive	Sensors and EMI shielding (e.g. strain sensing, self-regulated heating elements, over-current protector)
10^{-6}–10^{-11}	Electrostatic, Dissipative	Anti-static materials (e.g. fuel tanks, anti-static storage containers, electronics connectors)
$< 10^{-11}$	Insulating	Insulator

Table 7.1. *Classification and applications of polymer nanocomposites according to their electrical conductivity. Table adapted from Pang* et al. *(2014)*

However, not all nanofillers are connected to the percolated network, as shown in the middle and right insets in Figure 7.1. This happens because nanofillers have to be close enough to each other so that the conduction of electricity is possible between any two of them. The maximum separation between nanofillers that allows the transfer of electrons between the two of them is in the order of a few nanometers (Li *et al.* 2007, Hu *et al.* 2008a, Simmons 1963). Thus, some nanofillers are wasted as they do not take part of the percolated network and, as a consequence, do not contribute to the conduction of electricity. Few attempts have been made in order to quantify the amount of nanofillers that actually contribute (or not) to the conduction of electricity in polymer nanocomposites.

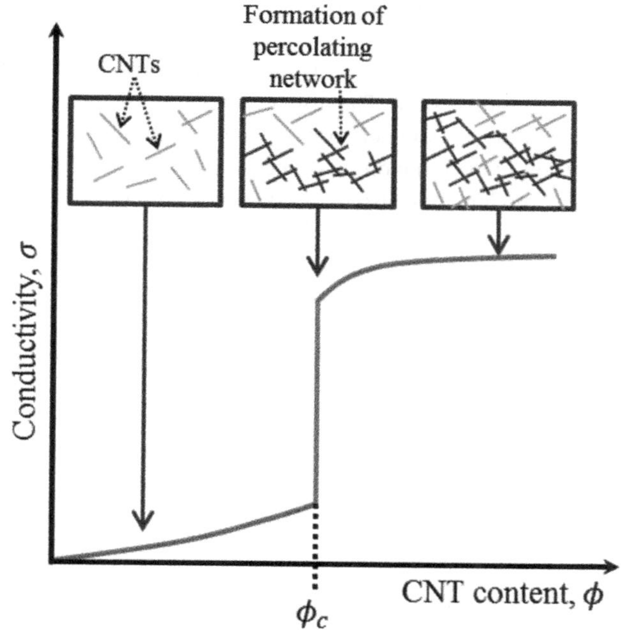

Figure 7.1. *Schematic representation of the electrical conductivity, σ, of a polymer composite as a function of CNT loading. A percolating network is formed at the percolation threshold, ϕ_c (adapted from Wernik and Meguid 2010)*

Computational models have been developed in order to study the electrical conductivity of polymers loaded with carbon-based nanofillers. Some focus on the study of the nanofiller network (Grujicic *et al.* 2004, Hu *et al.* 2008b, Mora *et al.* 2018b), while others use micromechanics models (Feng and Jiang 2013, Seidel and Puydupin-Jamin 2011) or the percolation theory (Foygel *et al.* 2005). Computational models that perform simulations of nanofiller networks are better suited to arrive at a definition of loading efficiency. Indeed, recently, a definition of loading efficiency based on the simulation of CNT networks was presented (Mora *et al.* 2018b). This definition can be used for other nanofillers (e.g., Mora *et al.* (2018a)). The study of the morphology of nanofiller networks may lead to a better understanding of the factors that affect the formation of a percolated network. In this way, guidelines can be defined such that a percolated network is formed using less nanofillers. Thus, polymer nanocomposites with higher electrical conductivity that require less nanofillers may be obtained.

Different strategies have been developed, from an experimental stand point, to improve the electrical conductivity of polymers. One such strategy involves mixing two nanofillers of different shapes, e.g., CNTs and GNPs (Al-Saleh 2016, Paszkiewicz *et al.* 2015, Chatterjee *et al.* 2012). Also, recently, hybrid particles have been developed in which CNTs are grown using another particle as a substrate, such as GNPs or silicon carbide (SiC). The use of these hybrid particles has shown promising results as hybrid particle networks are more conductive than those formed by single or mixed (non-hybrid) nanofillers (Dichiara *et al.* 2012, Zhao 2015). In this regard, computational models may also be used as a design tool so that the geometry of the hybrid particle that results in higher electrical conductivity is found.

In this chapter, carbon nanofiller networks will be studied to improve the conductivity of polymers using a computational approach. First, only CNT networks are studied to present a definition of loading efficiency. Then, the loading efficiency and electrical conductivity of CNT/polymer composites are studied. Later, hybrid-particle networks are studied. The definition of loading efficiency is also used for hybrid-particle networks and its relation with electrical conductivity is studied. Finally, the geometry of the hybrid particle is optimized and conclusions are presented.

7.2. Modeling and simulation of CNT/polymer nanocomposites

Different methodologies have been developed to study the material properties of polymers loaded with CNTs. Also, different methodologies have been followed to reproduce a geometric representation of a CNT network. Thus, an overview of these methodologies for geometric representation and modeling of electrical conductivity are presented in the following sections.

7.2.1. *Geometrical modeling of CNT/polymer nanocomposites*

Computational models have been developed to study CNT networks using, 2D and 3D models. In the case of 2D models, CNTs may be represented as straight (Rahman and Servati 2012, Zeng *et al.* 2011) or curved (Langley *et al.* 2018, Lin *et al.* 2010) lines, rectangles (Natsuki *et al.* 2005) or polygons (Li and Chou 2008). CNTs have been observed to curl inside polymer matrices, thus the usage of polygons is able to capture this experimentally observed behavior.

There are also quasi-3D models that consist of stacked layers of 2D models (Behnam *et al.* 2007, Jack *et al.* 2010). Another type of quasi-3D model consists of stacked layers of very thin 3D models. For instance, each layer of the model presented by Li *et al.* (2007) has a thickness of two inter-tube layers.

In the case of 3D models, some approaches have modeled CNTs as capped (Berhan and Sastry 2007a, Foygel *et al.* 2005, Grujicic *et al.* 2004) or uncapped (Bao *et al.* 2012, Gong *et al.* 2014, Hu *et al.* 2008a, Ounaies *et al.* 2003, Yu *et al.* 2010) straight cylinders. In order to capture the curliness of CNTs, they have been modeled using helical curves (Berhan and Sastry 2007b) or a sequence of n straight segments (Hu *et al.* 2008b, Lu *et al.* 2010, Mora *et al.* 2018b). However, helical CNTs are not representative of CNTs randomly dispersed in a polymer matrix. On the other hand, by allowing each of the n straight segments to have different orientations, a CNT may curl in a random manner. This type of representation has been shown to be representative of experimentally observed CNT networks (Han *et al.* 2014).

7.2.2. *Analysis of electrical conductivity*

Following the percolation theory, the electrical conductivity of a CNT/polymer composite can be calculated as:

$$\sigma = \sigma_0(\phi - \phi_c)^t, \qquad [7.1]$$

where t is known as the critical exponent and has a universal value of $t = 1.3$ for 2D models and $t = 2$ for 3D models (Stauffer and Aharony 1994), ϕ is the CNT content, ϕ_c is the percolation threshold and σ_0 depends on several factors, such as the CNT conductivity, contact resistance, network morphology and is sometimes considered to be the conductivity of the CNT (Foygel *et al.* 2005, Ounaies *et al.* 2003, Yu *et al.* 2010).

However, equation [7.1] is valid close to and above ϕ_c, but its range of validity has not been properly defined. Also, values of up to $t \approx 10$ have been found from experiments (Bauhofer and Kovacs 2009, Mutiso and Winey 2015), which are well above the universal values. This deviation from the universal values is not well understood yet and is attributed to a non-homogeneous distribution of CNTs, their type (metallic or semiconducting) or the hopping and tunneling of electrons (Mutiso and Winey 2015, Li *et al.* 2010, Yazdani *et al.* 2016, Thakre *et al.* 2009).

Given the limitations of percolation theory, other approaches to study electrical conductivity of CNT/polymer composites have been followed. There are micromechanics models, in which an equivalent conductivity is found through a homogenization process. Seidel and Lagoudas (2009) developed a micromechanics model based on the Mori–Tanaka homogenization method. In that model, they consider the CNT and a surrounding interphase as an equivalent filler. Then the equivalent conductivity of a matrix with this equivalent filler is obtained. García-Macías *et al.* (2017) also developed a micromechanics model based on the Mori–Tanaka method, but included effects of CNT agglomeration and curliness. Micromechanics models have also been used in combination with 2D and 3D finite element simulations (Ren and Seidel 2013, Lu 2017). In these type of models, the electrical conductivity can be obtained at different points in the material.

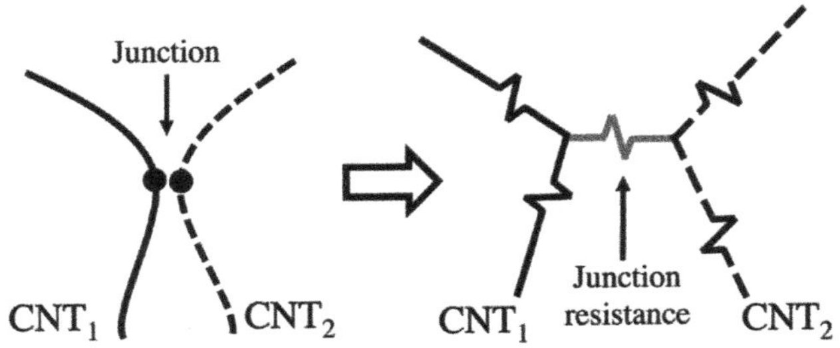

Figure 7.2. *Two CNTs sharing in electrical contact (left) and their resistors network (right)*

NOTE.– Lu (2017) performed simulation of GNP/polymer composites. However, their work is mentioned as the methodology can be applied to CNT/polymer composites.

Micromechanics models have resulted in good agreement with experimental results. However, they do not always consider that not all CNTs actively participate in the conduction of electricity in the composite (Seidel and Lagoudas 2009, Haj-Ali *et al.* 2014).

Finally, there are models that turn a CNT network into a resistor network to obtain the conductivity of the composite. In these models, two types of resistors are considered: CNT resistors and junction resistors. A CNT resistor comes from the intrinsic resistance of a CNT. A junction resistor comes from the flow of electrons between CNTs due to tunneling and hopping of electrons. Figure 7.2 shows an example of two CNTs and a junction and their corresponding resistor network. Panozzo *et al.* (2017) considered that CNTs form several parallel conductive paths, where each of these paths consists of several junction and CNT resistors connected in series. Thus, each parallel conductive path results in a parallel resistor. Then the conductivity of the composite is given by first calculating the equivalent resistance of such parallel resistors. However, Lubineau *et al.* (2017) found that, while there might be parallel paths for CNT loadings close to percolation threshold, as CNT content is increased all conductive paths tend to group into one large network. In other works, full random CNT networks are generated computationally and then turned into a random network of resistors (Hu *et al.* 2010, Li *et al.* 2007, Mora 2018). These types of models work following the same general steps:

– generate a random network of CNTs;

– find the percolated network. This is usually done using the Hoshen–Kopelman algorithm (Babalievski 1998, Hoshen and Kopelman 1976);

– extract the backbone from the percolated network. The backbone is the subset of a percolated network that actually carries electrical current. That is, the network that results after removing dead branches (open loops that carry no electricity) from the percolated network (such as the ones highlighted in red in Figure 7.3). This can be done using the direct electrifying algorithm (Li and Chou 2007, 2009b);

– turn the backbone into a network of resistors and, after applying a voltage, solve the equations associated with the resistor network to obtain the current passing through the composite;

– using Ohm's law, calculate the equivalent resistance of the composite. Using the composite's geometry, calculate its electrical conductivity (or resistivity).

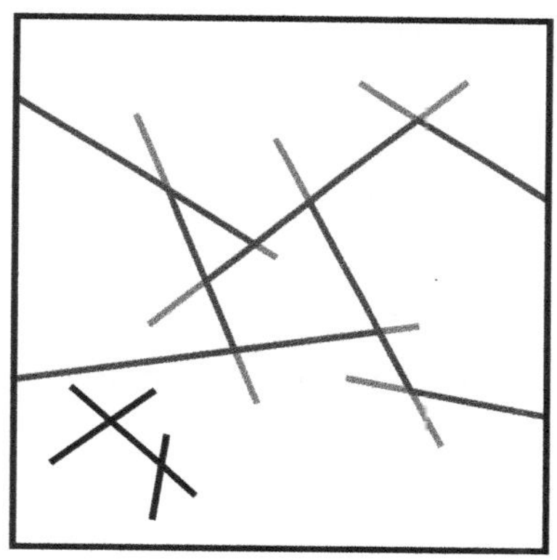

Figure 7.3. *Example of a random CNT network where the percolated network is highlighted in red and dead branches (open loops that carry no electricity) are highlighted in light gray. A non-percolated network is also shown in black. For a color version of this figure, see www.iste.co.uk/bai/nanocomposites.zip*

The advantage of using a resistor network approach lies in fact that the amount of non-percolated CNTs may be approximated directly from a CNT network.

7.3. Improving electrical conductivity of polymers loaded with CNTs

To provide guidelines to improve the electrical conductivity of CNT/polymer composites, first a definition of loading efficiency is presented. Then, its relationship with electrical conductivity and the parameters that affect these two quantities are studied.

7.3.1. *Introducing a definition of loading efficiency*

Mora *et al.* (2018b) introduced a definition of loading efficiency, η, as the fraction of CNTs that form part of the backbone network. In this way, loading efficiency is defined as:

$$\eta = \frac{V_{CNT,b}}{V_{CNT,T}} \quad [7.2]$$

where $V_{CNT,b}$ is the volume of CNTs in the backbone and $V_{CNT,T}$ is the total volume of CNTs inside a composite. Given the difficulty of measuring $V_{CNT,b}$, by fitting results from their simulations, Mora *et al.* (2018b) presented the following approximation for loading efficiency:

$$\eta\left(\phi_{red}\right) = \begin{cases} 0, & 0 \leq \phi < \phi_c \\ -0.1333\phi_{red}^2 + 0.5\phi_{red} + 0.0333, & \phi_c \leq \phi \end{cases} \quad [7.3]$$

where $\phi_{red} = \phi/\phi_c - 1$ is the reduced CNT volume fraction.

NOTE.– Although the approximation given in equation [7.3] fits the results from Mora *et al.* (2018b) well, it has some limitations. First, η has a maximum at $\phi_{red} = 1.8755$, after which loading efficiency starts decreasing. This is, of course, not expected to happen in actual composites. Additionally, $\eta < 0$ for $\phi_{red} > 3.8164$. Thus, the approximation of η in equation [7.3] is not recommended for $\phi_{red} > 1.8755$.

To overcome the limitations of approximation [7.3], Mora *et al.* (2020) fitted the simulation results from Mora *et al.* (2018b) using an exponential function as:

$$\eta = \begin{cases} 0; & 0 \leq \phi < \phi_c \\ -0.9431\exp\left(-0.4477\phi_{red}\right) + 1; & \phi_c \leq \phi \leq 1. \end{cases} \quad [7.4]$$

From equation [7.4], it is clear that as $\phi_{red} \to \infty$ we have that $\eta \to 1$, which is the maximum value of η obtained when all CNTs belong to the backbone.

It is noted that equation [7.2] (and its approximations in equations [7.3] and [7.4]) is the first definition for loading efficiency presented in the literature. However, Deng and Zheng (2008) presented an expression for the fraction of percolated CNTs, ξ, as:

$$\xi\left(\phi\right) = \begin{cases} 0, & 0 \leq \phi < \phi_c \\ \dfrac{\phi^{1/3} - \phi_c^{1/3}}{1 - \phi_c^{1/3}}, & \phi_c \leq \phi \leq 1. \end{cases} \quad [7.5]$$

This expression has been used in several research articles (e.g. Takeda *et al.* (2011), Ma *et al.* (2015), García-Macías *et al.* (2018)), but in no case has it been referred to as loading efficiency. In addition, its validity and limitations have not been studied.

Given that equations [7.3]–[7.5] approximate the amount of percolated CNTs, they are compared. First, it is observed that equation [7.5] is continuous for $0 \leq \phi \leq 1$, while equations [7.3] and [7.4] both have a discontinuity at $\eta(\phi_c)$. On the one hand, equation [7.5] is continuous since $\xi(\phi_c) = 0$, which is a non-realistic result as it implies no CNTs are in the backbone of a percolated network. On the other hand, the discontinuities of equations [7.3] and [7.4] at $\eta(\phi_c)$ reflect the sudden increase in electrical conductivity observed at a CNT concentration of ϕ_c (e.g. as suggested by the schematic in Figure 7.1).

Figure 7.4 shows ξ and η for three aspect ratios ($A_r = 100, 500, 1000$), where the percolation thresholds are taken from Mora *et al.* (2018b). Since both equations [7.3] and [7.4] approximate equation [7.2] using the same simulation results, for simplicity only one approximation of η is shown in Figure 7.4. Here, equation [7.4] is preferred as it does not have the limitations of equation [7.3]. As noted by Mora *et al.* (2018b), CNT aspect ratio has no effect over η and thus only one line is plotted in Figure 7.4. On the other hand, the differences in ξ become larger as the CNT aspect ratio is increased. Another important difference is that the three curves for ξ are well below the curve for η. Since approximation 7.4 is obtained from simulations of CNT networks, this suggests equation [7.5] under-estimates the fraction of percolated CNTs.

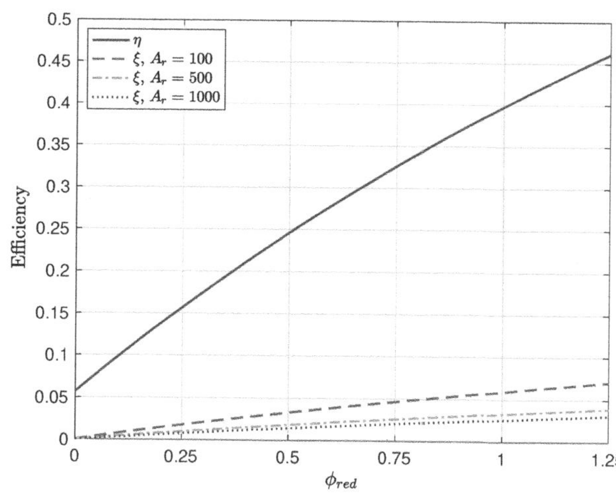

Figure 7.4. *Comparison of the fraction of percolated CNTs , ξ, as given by Deng and Zheng (2008), with the approximation of loading efficiency, η, given by Mora* et al. *(2020). For a color version of this figure, see www.iste.co.uk/bai/nanocomposites.zip*

From the observations above, the definition of loading efficiency presented by Mora *et al.* (2018b) is more representative of actual CNT networks and percolation behavior. Thus, the definition presented in equation [7.2] is recommended.

7.3.2. *Efficiency and electrical conductivity of polymers loaded with CNTs*

Now, the loading efficiency and electrical conductivity of CNT/polymer nanocomposites are compared. Figure 7.5(a) shows the loading efficiency obtained from simulations of CNT networks for three aspect ratios, A_r, of 100, 500 and 1000. It is clear from that figure that, for a given CNT content, efficiency is higher for polymers loaded with CNTs with larger aspect ratio. It is also observed that loading efficiency increases faster as the aspect ratio is increased.

Figure 7.5(b) shows the electrical conductivity corresponding to the efficiencies shown in Figure 7.5(a). First, it is observed that an increase in efficiency corresponds to an increase in electrical conductivity. Also, for a given CNT content, electrical conductivity is higher for polymers loaded with CNTs with a larger aspect ratio, which is the same behavior observed with loading efficiency. However, the differences in electrical conductivity for composites with CNTs of different aspect ratios are reduced as CNT content is increased.

Thus, in principle, the electrical conductivity of composites may be improved by using CNTs with a large aspect ratio. However, this benefit depends on the CNT concentration. For instance, Figure 7.5(b) suggests that, for CNT contents of around 0.3 % volume fraction and above, electrical conductivity does not change much if the aspect ratio of CNTs is increased above 500. In addition, CNTs with a large aspect ratio commonly form agglomerations (Brigandi *et al.* 2017, Mierczynska *et al.* 2007, Pan *et al.* 2010, Xin and Li 2011) that reduce the percolation effect and might even act as defects (Truong *et al.* 2007). This is to be taken into consideration in applications where mechanical deformation is also required (e.g. strain sensors).

7.3.3. *Influence of junction resistance*

An important parameter in calculating the electrical conductivity of CNT/polymer composites is the junction resistance. To approximate the junction resistance, many efforts (e.g. Bao *et al.* (2011), Hu *et al.* (2012), Li *et al.* (2007), Hu *et al.* (2008a), García-Macías *et al.* (2017)) have relied on the theory developed by Simmons (1963). For instance, Hu *et al.* (2008a) presented the following approximation for the junction resistance:

$$R_J = \frac{h^2 t}{Ae^2\sqrt{2m\lambda}} \exp\left(\frac{4\pi t}{h}\sqrt{2m\lambda}\right), \qquad [7.6]$$

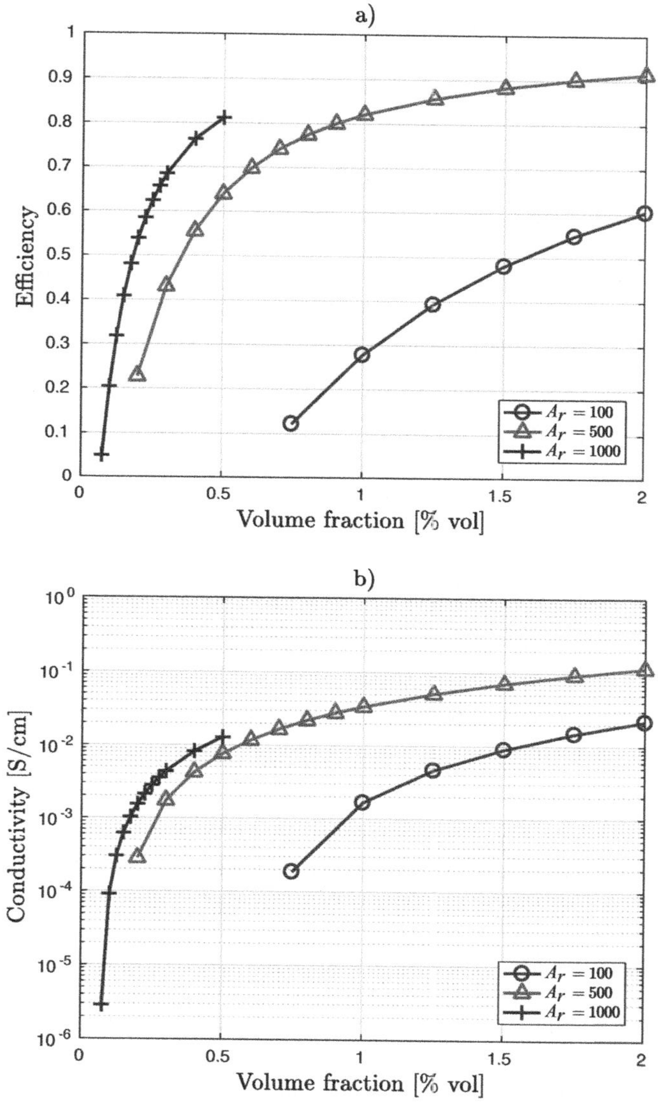

Figure 7.5. *Comparison of efficiency (a) and electrical conductivity (b) as a function of volume fraction for different aspect ratios (A_r) (Mora* et al. *2018b). For a color version of this figure, see www.iste.co.uk/bai/nanocomposites.zip*

where h is Planck's constant ($6.62606957 \times 10^{-34}$ kg·m^2/s), t is the thickness of the polymer layer separating two CNTs, A is the cross-sectional area of the tunnel, λ is the height of the barrier, and e and m are the charge ($1.602176565 \times 10^{-19}$ C) and mass of an electron ($9.10938291 \times 10^{-31}$ kg), respectively.

However, the actual value of the junction resistance is dependent on several factors, such as the nanotube type and diameter, junction area and electrical properties of the polymer matrix (Li and Chou 2009a, Mutiso and Winey 2015, Buldum and Lu 2001, Fuhrer 2000). In contrast, the approximation of equation [7.6] only considers the separation of the CNTs and the properties of the matrix (through the height barrier). In addition, the effect of the junction resistance on the electrical conductivity of the composite has been rarely investigated (Li *et al.* 2007, De Vivo *et al.* 2013).

A study of the effect of junction resistance on the composite's electrical conductivity using a computational approach is presented in Figure 7.5. In this figure, each electrical conductivity curve is obtained using a constant value for the junction resistance for all junctions in the resistor network. It is observed that for values of R_J of 10^6 Ω or below there is little effect on further reducing the junction resistance. On the other hand, when $R_J \geq 10^7$, reducing the junction resistance in one order of magnitude results in a similar increase in the conductivity of the composite. These results indicate the benefit that can be obtained from research dealing with reducing the effects of junction resistance (Zhou and Lubineau 2013, Hermant *et al.* 2009, Ventura *et al.* 2015). Conversely, the results in Figure 7.5 suggest that there is a limit above which reducing the effect of junction resistance does not considerably increase the conductivity of the composite.

7.4. Improving electrical conductivity of polymers loaded with hybrid particles

7.4.1. *Hybrid particles*

Up until this point, the focus has been on polymers loaded with CNTs to increase their electrical conductivity. However, a recent strategy to improve the electrical conductivity of polymers consists of using nanofillers of different shapes to create new (hybrid) particles (He *et al.* 2017, Li *et al.* 2014a). In this way, CNTs have been grown using other nanofillers as substrates, such as GNPs, silicon carbide (SiC) and alumina (Al_2O_3) (Li *et al.* 2014b, 2013b). An example of such particles, which consists of CNTs grown on a GNP, is shown in the schematic in Figure 7.7. For the rest of the chapter, we will focus on hybrid particles consisting of CNTs grown on GNPs as substrates (CNT-GNP hybrid particles).

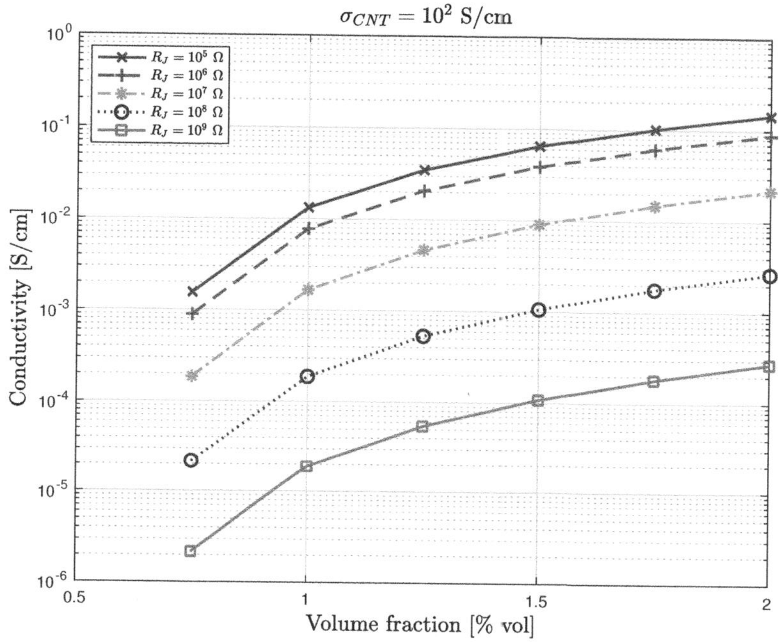

Figure 7.6. *Sensitivity analysis of junction resistance (Mora* et al. *2018b). For a color version of this figure, see www.iste.co.uk/bai/nanocomposites.zip*

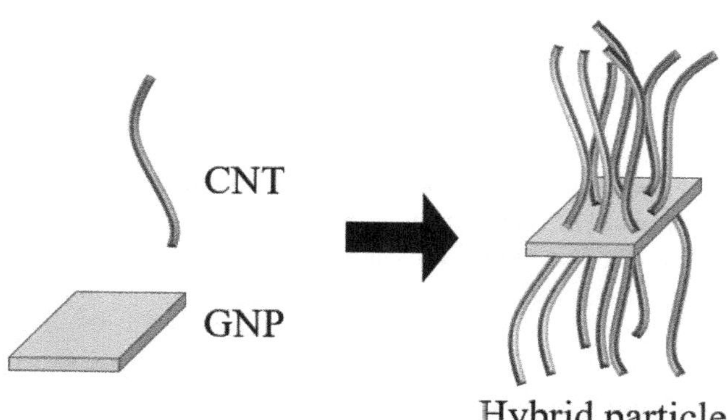

Figure 7.7. *Schematic of a hybrid particle consisting of CNTs grown on a GNP*

Experiments have shown that polymers loaded with CNT-GNP hybrid particles are more electrically conductive than polymers loaded with only CNTs, only GNPs, or a mixture of CNTs and GNPs (Dichiara *et al.* 2012, Zhao 2015). In addition, polymers loaded with CNT-GNP hybrid particles have shown potential as strain sensors due to their piezoresistive behavior (Zhao and Bai 2015). The high electrical conductivity of polymers loaded with CNT-GNP hybrid particles has been attributed to the GNPs acting as CNT concentrators (Zhao and Bai 2015). Here, the CNTs of a hybrid particle form a bundle of connected CNTs. Thus, when one CNT of the bundle is connected to a percolated network, all of the CNTs in the bundle are also connected. This description is similar to that of a segregated structure, in which nanoparticles are constrained to certain locations in a composite. Indeed, by using a computational tool, it was shown that hybrid particle networks actually form segregated structures (Mora *et al.* 2018a).

7.4.2. *Efficiency and electrical conductivity of CNT-GNP hybrid-particle networks*

Following the definition in equation [7.2], the loading efficiency of polymers loaded with CNT-GNP hybrid particles is calculated by performing simulations of the networks they form inside polymers. Loading efficiency for CNT-GNP hybrid particles is shown in Figure 7.8(a) for different volume fractions of carbon content. For comparison, the loading efficiencies of CNTs, GNPs and mixed CNTs and GNPs are also shown in the figure. For the four types of nanofillers, loading efficiency increases by increasing carbon content. The increase in efficiency is faster for low loadings and, as the carbon content is increased, efficiency increases at a lower rate. This is due to large changes in the morphology of the nanofiller network occurring at loadings close to the percolation threshold (Lubineau *et al.* 2017).

From Figure 7.8(a), it is also observed that hybrid particles show the highest loading efficiency for all volume fractions. The high efficiency shown by the hybrid particles is a direct result of them forming a segregated structure (Mora *et al.* 2018a). Since CNTs are attached at GNPs, they are close to each other and it is easier for them to be in contact with each other. Thus, the more CNTs are in the backbone, the higher the efficiency.

The increased loading efficiency of CNT-GNP hybrid particles is also reflected in increased electrical conductivity compared to the other nanofillers, as shown in Figure 7.8(b). Although the differences in electrical conductivity are not as clear as with efficiency, hybrid particles show the highest electrical conductivity for all loadings. The difference in electrical conductivity is especially clear at low loadings, that is, close to the percolation threshold. Thus, applications that require loadings close to the percolation threshold may benefit from using polymers loaded with CNT-GNP hybrid particles.

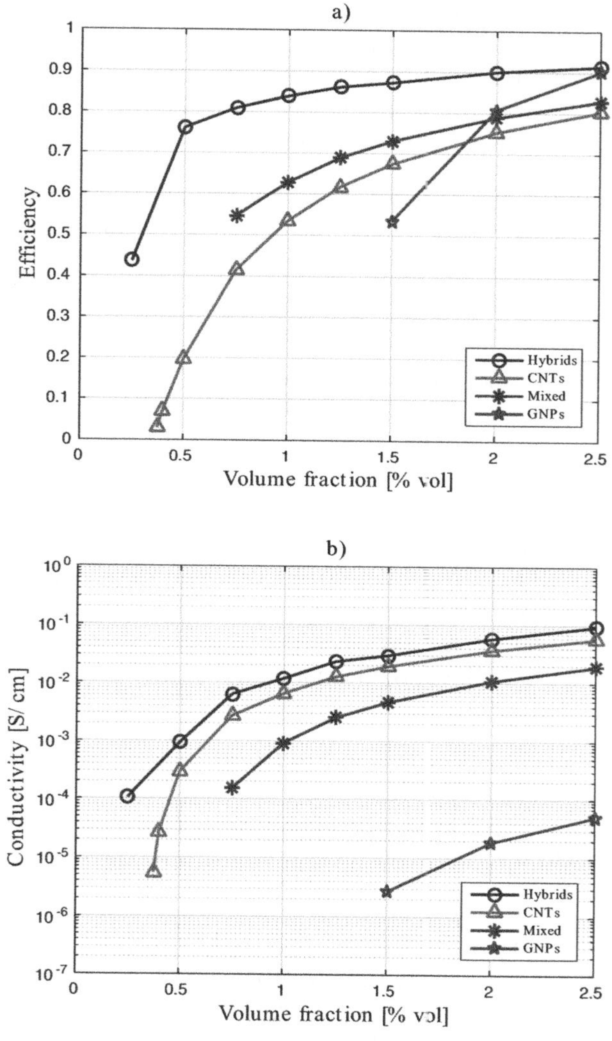

Figure 7.8. *a) Nanofiller efficiencies at different volume fractions of carbon content. b) Electrical conductivities for different nanofillers at different volume fractions of carbon content (Mora* et al. *2018a). For a color version of this figure, see www.iste.co.uk/bai/nanocomposites.zip*

7.4.3. *Optimization of the hybrid-particle geometry*

Finally, the use of a computational tool to optimize the geometry of the CNT-GNP hybrid particle is shown here. Following Mora *et al.* (2018a), the optimization parameters are the CNT aspect ratio, $A_r = l_{CNT}/r_{CNT}$, the GNP inverse aspect ratio, $A_{ri} = t_{GNP}/l_{GNP}$, and the number of CNTs per micron squared on GNPs, N. Here, l_{CNT} and r_{CNT} are the CNT length and radius, respectively, and t_{GNP} and l_{GNP} are the GNP thickness and side length. Here, a GNP is modeled as parallelepiped with a square base.

Based on typical geometries of hybrid particles grown experimentally (Li *et al.* 2013a, Zhao and Bai 2015, Zhao 2015), the search space for the optimization is defined as follows:

– for A_r, the CNT radius is kept constant at $r_{CNT} = 5$ nm, while its length varies within $3 \leq l_{CNT} \leq 8$ μm. Thus $A_r \in [300, 800]$;

– for A_{ri}, the GNP side length is kept constant at $l_{GNP} = 3$ μm, while its thickness varies within $10 \leq t_{GNP} \leq 60$ nm. Thus $A_{ri} \in [1/300, 6/300]$;

– N varies in the range $[10, 100]$.

NOTE.– Usually the aspect ratio is defined as the ratio of the larger dimension to the smaller dimension. This results in the GNP aspect ratio usually being defined as $A_r = l_{GNP}/t_{GNP}$ (Safdari and Al-Haik 2012, Chen *et al.* 2003, Li and Zhong 2011). However, when studying the effect of modifying the larger dimension (e.g. l_{GNP} for GNPs), it is more useful to take this definition. Since t_{GNP} is the variable that is modified in the optimization process, the ratio of the smaller dimension to the larger is considered here.

The composites for the simulations are taken to be cubes of 30 μm per side, while the carbon content is kept constant at 1% volume fraction. The optimization is performed by keeping one of the three optimization parameters constant and varying the other two within their specified ranges. The results of this procedure are shown in Figures 7.9–7.11.

Figure 7.9 shows a 3D plot of the conductivity when a value of $A_{ri} = 3/300$ is kept constant. The insets in Figure 7.9 illustrate the changes in the microstructure as A_r and N are increased. Since the carbon content is kept constant, increasing A_r (N) results in having less hybrid particles but with longer CNTs (more CNTs per μm^2). These changes are indicated with the arrows connecting the insets in Figure 7.9. It is observed in Figure 7.9 that increasing either A_r or N results in an increase in the electrical conductivity of the composite. A similar result is observed in polymers loaded with CNTs, where increasing their aspect ratio results in composites with higher electrical conductivity (Mutiso and Winey 2015, Li *et al.* 2008b). It is also observed that, for $N > 55$ CNTs/μm^2, there is no significant increase in the composite's conductivity. Thus, this value can be used as an upper limit when synthesizing hybrid particles.

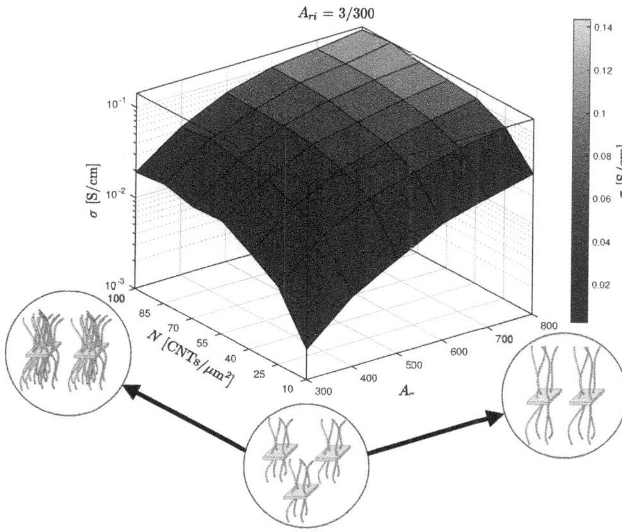

Figure 7.9. *Electrical conductivity as a function of CNT aspect ratio, A_r, and CNT density, N. The GNP inverse aspect ratio is kept constant at $A_{ri} = 3/300$. Insets show changes in the microstructure when varying N or A_r in the directions indicated by the arrows. For a color version of this figure, see www.iste.co.uk/bai/nanocomposites.zip*

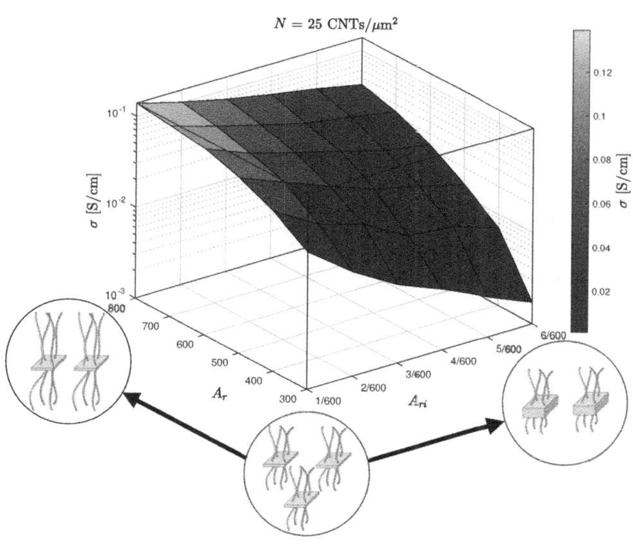

Figure 7.10. *Electrical conductivity as a function of CNT aspect ratio, A_r, and GNP inverse aspect ratio, A_{ri}. The CNT density was kept constant at $N = 25$ CNTs/μm^2. Insets show changes in the microstructure when varying A_r or A_{ri} in the directions indicated by the arrows. For a color version of this figure, see www.iste.co.uk/bai/nanocomposites.zip*

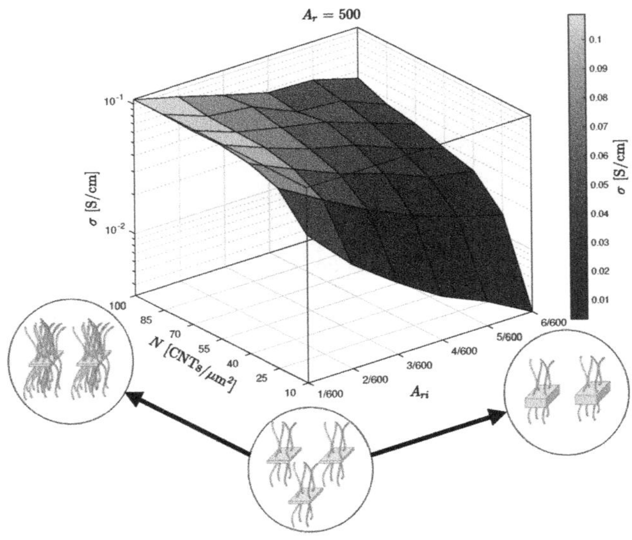

Figure 7.11. *Electrical conductivity as a function of GNP inverse aspect ratio, A_{ri}, and CNT density, N. The CNT aspect ratio was kept constant at $A_r = 500$. Insets show changes in the microstructure when varying N or A_{ri} in the directions indicated by the arrows. For a color version of this figure, see www.iste.co.uk/bai/nanocomposites.zip*

Figure 7.10 shows a 3D plot of the conductivity when a value of $N = 25$ CNTs/μm^2 is kept constant. The insets in Figure 7.9 illustrate the changes in the microstructure, where increasing A_r (A_{ri}) results in having less hybrid particles but with longer CNTs (thicker GNPs). It is observed that increasing A_r but decreasing A_{ri} results in an increase in the electrical conductivity of the composite. That is, higher electrical conductivities can be achieved when most of the carbon content is made up of CNTs. In the limiting case, tying up a bundle of CNTs of length $2l_{CNT}$ by the middle would result in the highest electrical conductivity.

NOTE.– In an experimental set up, as the GNPs get thinner, they might not be able to maintain the GNPs on their surfaces. Thus, experiments are needed to assess the minimum thickness that guarantees the integrity of the hybrid particles.

Figure 7.11 shows a 3D plot of the conductivity when a value of $A_r = 500$ is kept constant. The insets in Figure 7.9 illustrate the changes in the microstructure, where increasing N (A_{ri}) results in having less hybrid particles but with more CNTs per μm^2 (thicker GNPs). Results are, of course, consistent with those observed in Figures 7.9 and 7.10. In that sense, the upper limit of 55 CNTs/μm^2 and thin GNPs are again drawn as recommendations to optimize the CNT-GNP hybrid particle geometry.

7.5. Conclusions

There are different strategies to study polymer composites loaded with carbon nanofillers using computational methods. Each method has its own benefits and shortcomings. In any case, computational methods can be used to provide guidelines that improve the usage of nanoparticles and further increase the electrical conductivity of polymer composites.

Computational tools can also be used to offer some insight where available tools and instruments can not. For instance, full CNT networks can be studied and their backbone extracted using computational tools. In this way, a definition of loading efficiency was proposed based on the backbone network. Given a definition of loading efficiency, its dependence on CNT geometry and content can be studied. It was found that increasing the CNT aspect ratio and content results in increasing loading efficiency. However, as the CNT content is increased, the dependence on aspect ratio is reduced. Thus, in principle, large aspect ratio CNTs are recommended to increase electrical conductivity due to their higher efficiency. However, the advantage of large aspect ratio CNTs is reduced as the CNT content is increased. It was also found that reducing the junction resistance in one order of magnitude may result in an improvement of one order of magnitude in the composite's conductivity. In this regard, research that aims to reduce the effects of the junction resistance is highlighted.

The networks formed by recently developed CNT-GNP hybrid particles were studied using a computational tool. It was found that these networks were more conductive due to their higher efficiency. The geometry of this hybrid particle was also optimized using a computational tool. It is suggested to synthesize hybrid particles with no more than 55 CNT/μm^2, as larger CNT densities do not significantly improve the electrical conductivity of the composite. It is also suggested that the majority of the carbon content in the hybrid particles is made up of CNTs.

Finally, the work presented in this chapter shows how computational tools and experiments may complement each other to further improve properties of materials.

7.6. References

Al-Saleh, M.H. (2016). Electrical, EMI shielding and tensile properties of PP/PE blends filled with GNP: CNT hybrid nanofiller. *Synthetic Metals*, 217, 322–330.

Babalievski, F. (1998). Cluster counting: The Hoshen-Kopelman algorithm vs. spanning tree approaches. *International Journal of Modern Physics C*, 9(1), 18.

Bao, W., Meguid, S., Zhu, Z., Meguid, M. (2011). Modeling electrical conductivities of nanocomposites with aligned carbon nanotubes. *Nanotechnology*, 22(48), 485704.

Bao, W., Meguid, S., Zhu, Z., Pan, Y., Weng, G. (2012). A novel approach to predict the electrical conductivity of multifunctional nanocomposites. *Mechanics of Materials*, 46, 129–138.

Bauhofer, W. and Kovacs, J.Z. (2009). A review and analysis of electrical percolation in carbon nanotube polymer composites. *Composites Science and Technology*, 69(10), 1486–1498.

Behnam, A., Guo, J., Ural, A. (2007). Effects of nanotube alignment and measurement direction on percolation resistivity in single-walled carbon nanotube films. *Journal of Applied Physics*, 102(4), 044313.

Berhan, L. and Sastry, A. (2007a). Modeling percolation in high-aspect-ratio fiber systems I. Soft-core versus hard-core models. *Physical Review E*, 75(4), 041120.

Berhan, L. and Sastry, A. (2007b). Modeling percolation in high-aspect-ratio fiber systems II. The effect of waviness on the percolation onset. *Physical Review E*, 75(4), 041121.

Brigandi, P.J., Cogen, J.M., Reffner, J.R., Wolf, C.A., Pearson, R.A. (2017). Influence of carbon black and carbon nanotubes on the conductivity, morphology, and rheology of conductive ternary polymer blends. *Polymer Engineering and Science Engineering*, 57(12), 1329–1339.

Buldum, A. and Lu, J.P. (2001). Contact resistance between carbon nanotubes. *Physical Review B*, 63(16), 161403.

Chatterjee, S., Nafezarefi, F., Tai, N.H., Schlagenhauf, L., Nüesch, F.A., Chu, B.T.T. (2012). Size and synergy effects of nanofiller hybrids including graphene nanoplatelets and carbon nanotubes in mechanical properties of epoxy composites. *Carbon*, 50(15), 5380–5386.

Chen, G., Weng, W., Wu, D., Wu, C. (2003). PMMA/graphite nanosheets composite and its conducting properties. *European Polymer Journal*, 39(12), 2329–2335.

De Vivo, B., Lamberti, P., Spinelli, G., Tucci, V. (2013). Numerical investigation on the influence factors of the electrical properties of carbon nanotubes-filled composites. *Journal of Applied Physics*, 113(24), 244301.

Deng, F. and Zheng, Q.-S. (2008). An analytical model of effective electrical conductivity of carbon nanotube composites. *Applied Physics Letters*, 92(7), 071902.

Dichiara, A., Yuan, J.K., Yao, S.H., Sylvestre, A., Bai, J.B. (2012). Chemical vapor deposition synthesis of carbon nanotube-graphene nanosheet hybrids and their application in polymer composites. *Journal of Nanoscience and Nanotechnology*, 12(9), 6935–6940.

Ezquerra, T., Martinez-Salazar, J., Baltá Calleja, F. (1986). Percolation threshold of conductive polycarbonate/carbon composites as revealed by electron microscopy. *Journal of Materials Science Letters*, 5(10), 1065–1066.

Feng, C. and Jiang, L. (2013). Micromechanics modeling of the electrical conductivity of carbon nanotube (CNT)-polymer nanocomposites. *Composites Part A: Applied Science and Manufacturing*, 47, 143–149.

Foygel, M., Morris, R., Anez, D., French, S., Sobolev, V. (2005). Theoretical and computational studies of carbon nanotube composites and suspensions: Electrical and thermal conductivity. *Physical Review B*, 71(10), 104201.

Fuhrer, M.S. (2000). Crossed nanotube junctions. *Science*, 288(5465), 494–497.

García-Macías, E., D'Alessandro, A., Castro-Triguero, R., Pérez-Mira, D., Ubertini, F. (2017). Micromechanics modeling of the electrical conductivity of carbon nanotube cement-matrix composites. *Composites Part B: Engineering*, 108, 451–469.

García-Macías, E., Rodríguez-Tembleque, L., Sáez, A., Ubertini, F. (2018). Crack detection and localization in RC beams through smart MWCNT/epoxy strip-like strain sensors. *Smart Materials and Structures*, 27(11), 115022.

Ge, C. and Cosgrove, K. (2015). Preparation of PVOH coatings with graphene nanoplatelets for electrostatic discharge protective packaging. *Journal of Electrostatics*, 77, 157–162.

Gong, S., Zhu, Z., Meguid, S. (2014). Carbon nanotube agglomeration effect on piezoresistivity of polymer nanocomposites. *Polymer*, 55(21), 5488–5499.

Grujicic, M., Cao, G., Roy, W. (2004). A computational analysis of the percolation threshold and the electrical conductivity of carbon nanotubes filled polymeric materials. *Journal of Materials Science*, 39(14), 4441–4449.

Haj-Ali, R., Zemer, H., El-Hajjar, R., Aboudi, J. (2014). Piezoresistive fiber-reinforced composites: A coupled nonlinear micromechanical-microelectrical modeling approach. *International Journal of Solids and Structures*, 51(2), 491–503.

Han, F., Azdoud, Y., Lubineau, G. (2014). Computational modeling of elastic properties of carbon nanotube/polymer composites with interphase regions. Part I: Micro-structural characterization and geometric modeling. *Computational Materials Science*, 81, 641–651.

He, D., Fan, B., Zhao, H., Yang, M., Wang, H., Bai, J., Li, W., Zhou, X., Bai, J. (2017). Multifunctional polymer composites reinforced by carbon nanotubes – Alumina hybrids with urchin-like structure. *Materials Today Communications*, 11, 94–102.

Hermant, M.C., Klumperman, B., Kyrylyuk, A.V., Van der Schoot, P., Koning, C.E. (2009). Lowering the percolation threshold of single-walled carbon nanotubes using polystyrene/poly(3,4-ethylenedioxythiophene): Poly(styrene sulfonate) blends. *Soft Matter*, 5(4), 878.

Hoshen, J. and Kopelman, R. (1976). Percolation and cluster distribution I. Cluster multiple labeling technique and critical concentration algorithm. *Physical Review B*, 14(8), 3438–3445.

Hu, N., Karube, Y., Yan, C., Masuda, Z., Fukunaga, H. (2008a). Tunneling effect in a polymer/carbon nanotube nanocomposite strain sensor. *Acta Materialia*, 56(13), 2929–2936.

Hu, N., Masuda, Z., Yan, C., Yamamoto, G., Fukunaga, H., Hashida, T. (2008b). The electrical properties of polymer nanocomposites with carbon nanotube fillers. *Nanotechnology*, 19(21), 215701.

Hu, N., Karube, Y., Arai, M., Watanabe, T., Yan, C., Li, Y., Liu, Y., Fukunaga, H. (2010). Investigation on sensitivity of a polymer/carbon nanotube composite strain sensor. *Carbon*, 48(3), 680–687.

Hu, B., Hu, N., Li, Y., Akagi, K., Yuan, W., Watanabe, T., Cai, Y. (2012). Multi-scale numerical simulations on piezoresistivity of CNT/polymer nanocomposites. *Nanoscale Research Letters*, 7, 402.

Jack, D.A., Yeh, C.-S., Liang, Z., Li, S., Park, J.G., Fielding, J.C. (2010). Electrical conductivity modeling and experimental study of densely packed SWCNT networks. *Nanotechnology*, 21(19), 195703.

Langley, D., Lagrange, M., Nguyen, N., Bellet, D. (2018). Percolation in networks of 1-dimensional objects: Comparison between Monte Carlo simulations and experimental observations. *Nanoscale Horizons*, 3(5), 545–550.

Li, C. and Chou, T.-W. (2007). A direct electrifying algorithm for backbone identification. *Journal of Physics A: Mathematical and Theoretical*, 40(49), 14679–14686.

Li, C. and Chou, T.-W. (2008). Modeling of damage sensing in fiber composites using carbon nanotube networks. *Composites Science and Technology*, 68(15/16), 3373–3379.

Li, C. and Chou, T.-W. (2009a). Electrical conductivities of composites with aligned carbon nanotubes. *Journal of Nanoscience and Nanotechnology*, 9(4), 2518–2524.

Li, C. and Chou, T.-W. (2009b). Precise determination of backbone structure and conductivity of 3D percolation networks by the direct electrifying algorithm. *International Journal of Modern Physics C*, 20(800), 422–433.

Li, B. and Zhong, W.H. (2011). Review on polymer/graphite nanoplatelet nanocomposites. *Journal of Materials Science*, 46(17), 5595–5614.

Li, C., Thostenson, E.T., Chou, T.-W. (2007). Dominant role of tunneling resistance in the electrical conductivity of carbon nanotube-based composites. *Applied Physics Letters*, 91(22), 223114.

Li, C., Thostenson, E.T., Chou, T.-W. (2008a). Effect of nanotube waviness on the electrical conductivity of carbon nanotube-based composites. *Composites Science and Technology*, 68(6), 1445–1452.

Li, C., Thostenson, E.T., Chou, T.-W. (2008b). Sensors and actuators based on carbon nanotubes and their composites: A review. *Composites Science and Technology*, 68(6), 1227–1249.

Li, C., Thostenson, E.T., Chou, T.-W. (2010). Carbon-nanotube-based composites and damage sensing. In *Multifunctional Polymer Nanocomposites*, Jinsong, L., Lau, A.K.T. (eds). CRC Press, Boca Raton.

Li, W., Dichiara, A., Bai, J. (2013a). Carbon nanotube-graphene nanoplatelet hybrids as high-performance multifunctional reinforcements in epoxy composites. *Composites Science and Technology*, 74, 221–227.

Li, W., He, D., Bai, J. (2013b). The influence of nano/micro hybrid structure on the mechanical and self-sensing properties of carbon nanotube-microparticle reinforced epoxy matrix composite. *Composites Part A: Applied Science and Manufacturing*, 54, 28–36.

Li, W., Dichiara, A., Zha, J., Su, Z., Bai, J. (2014a). On improvement of mechanical and thermo-mechanical properties of glass fabric/epoxy composites by incorporating CNT-Al2O3 hybrids. *Composites Science and Technology*, 103, 36–43.

Li, W., He, D., Dang, Z., Bai, J. (2014b). *In situ* damage sensing in the glass fabric reinforced epoxy composites containing CNT-Al2O3 hybrids. *Composites Science and Technology*, 99, 8–14.

Lin, C., Wang, H., Yang, W. (2010). Variable percolation threshold of composites with fiber fillers under compression. *Journal of Applied Physics*, 108(1), 013509-1–013509-5.

Lu, X. (2017). Multiscale electro-mechanical modeling of graphene/polymer nanocomposites. PhD Thesis, Université Paris-Saclay, Gif-sur-Yvette.

Lu, W., Chou, T.-W., Thostenson, E.T. (2010). A three-dimensional model of electrical percolation thresholds in carbon nanotube-based composites. *Applied Physics Letters*, 96(22), 223106.

Lubineau, G., Mora, A., Han, F., Odeh, I.N., Yaldiz, R. (2017). A morphological investigation of conductive networks in polymers loaded with carbon nanotubes. *Computational Materials Science*, 130, 21–38.

Ma, X., Scarpa, F., Peng, H.X., Allegri, G., Yuan, J., Ciobanu, R. (2015). Design of a hybrid carbon fibre/carbon nanotube composite for enhanced lightning strike resistance. *Aerospace Science and Technology*, 47, 367–377.

Mierczynska, A., Mayne-L'Hermite, M., Boiteux, G., Jeszka, J.K. (2007). Electrical and mechanical properties of carbon nanotube/ultrahigh-molecular-weight polyethylene composites prepared by a filler prelocalization method. *Journal of Applied Polymer Science*, 105, 158–168.

Mora, A. (2018). A morphology study of nanofiller networks in polymer nano-composites: Improving their electrical conductivity through better doping strategies. PhD Thesis, Université des sciences et technologies du roi Abdallah, Djeddah.

Mora, A., Han, F., Lubineau, G. (2018a). Computational modeling of electrically conductive networks formed by graphene nanoplatelet-carbon nanotube hybrid particles. *Modelling and Simulation in Materials Science and Engineering*, 26(3), 035010.

Mora, A., Han, F., Lubineau, G. (2018b). Estimating and understanding the efficiency of nanoparticles in enhancing the conductivity of carbon nanotube/polymer composites. *Results in Physics*, 10, 81–90.

Mora, A., Verma, P., Kumar, S. (2020). Electrical conductivity of CNT/polymer composites: 3D printing, measurements and modeling. *Composites Part B: Engineering*, 183, 107600.

Mutiso, R.M. and Winey, K.I. (2015). Electrical properties of polymer nanocomposites containing rod-like nanofillers. *Progress in Polymer Science*, 40, 63–84.

Nanni, F., Mayoral, B.L., Madau, F., Montesperelli, G., McNally, T. (2012). Effect of MWCNT alignment on mechanical and self-monitoring properties of extruded PET-MWCNT nanocomposites. *Composites Science and Technology*, 72(10), 1140–1146.

Natsuki, T., Endo, M., Takahashi, T. (2005). Percolation study of orientated short-fiber composites by a continuum model. *Physica A: Statistical Mechanics and its Applications*, 352(2–4), 498–508.

Ounaies, Z., Park, C., Wise, K., Siochi, E., Harrison, J. (2003). Electrical properties of single wall carbon nanotube reinforced polyimide composites. *Composites Science and Technology*, 63(11), 1637–1646.

Pan, Y., Li, L., Chan, S.H., Zhao, J. (2010). Correlation between dispersion state and electrical conductivity of MWCNTs/PP composites prepared by melt blending. *Composites Part A: Applied Science and Manufacturing*, 41(3), 419–426.

Pang, H., Xu, L., Yan, D.X., Li, Z.M. (2014). Conductive polymer composites with segregated structures. *Progress in Polymer Science*, 39(11), 1908–1933.

Panozzo, F., Zappalorto, M., Quaresimin, M. (2017). Analytical model for the prediction of the piezoresistive behavior of CNT modified polymers. *Composites Part B: Engineering*, 109, 53–63.

Paszkiewicz, S., Szymczyk, A., Sui, X.M., Wagner, H.D., Linares, A., Ezquerra, T.A., Rosaniec, Z. (2015). Synergetic effect of single-walled carbon nanotubes (SWCNT) and graphene nanoplatelets (GNP) in electrically conductive PTT-block-PTMO hybrid nanocomposites prepared by in situ polymerization. *Composites Science and Technology*, 118, 72–77.

Rahman, R. and Servati, P. (2012). Effects of inter-tube distance and alignment on tunnelling resistance and strain sensitivity of nanotube/polymer composite films. *Nanotechnology*, 23(5), 055703.

Ren, X. and Seidel, G. (2013). Computational micromechanics modeling of piezoresistivity in carbon nanotube-polymer nanocomposites. *Composite Interfaces*, 20(9), 693–720.

Safdari, M. and Al-Haik, M. (2012). Electrical conductivity of synergistically hybridized nanocomposites based on graphite nanoplatelets and carbon nanotubes. *Nanotechnology*, 23(40), 405202.

Seidel, G. and Lagoudas, D. (2009). A micromechanics model for the electrical conductivity of nanotube-polymer nanocomposites. *Journal of Composite Materials*, 43(9), 917–941.

Seidel, G. and Puydupin-Jamin, A. (2011). Analysis of clustering, interphase region, and orientation effects on the electrical conductivity of carbon nanotube-polymer nanocomposites via computational micromechanics. *Mechanics of Materials*, 43(12), 755–774.

Simmons, J.G. (1963). Generalized formula for the electric tunnel effect between similar electrodes separated by a thin insulating film. *Journal of Applied Physics*, 34(1963), 1793–1803.

Stauffer, D. and Aharony, A. (1994). *Introduction to Percolation Theory*, 2nd edition. Taylor & Francis, Abingdon-on-Thames.

Takeda, T., Shindo, Y., Kuronuma, Y., Narita, F. (2011). Modeling and characterization of the electrical conductivity of carbon nanotube-based polymer composites. *Polymer*, 52(17), 3852–3856.

Thakre, P.R., Bisrat, Y., Lagoudas, D.C. (2009). Electrical and mechanical properties of carbon nanotube-epoxy nanocomposites. *Journal of Applied Polymer Science*, 116, 191–202.

Truong, V.T., Tsang, K.M.C., Keough, S.J., St John, N.A. (2006). Effect of sonication on the mechanical properties of poly (vinyl alcohol)/carbon nanotube composites. *Micro- and Nanotechnology: Materials, Processes, Packaging, and Systems III*, Proc. SPIE 6415, 641503.

Ventura, I.A., Zhou, J., Lubineau, G. (2015). Investigating the inter-tube conduction mechanism in polycarbonate nanocomposites prepared with conductive polymer-coated carbon nanotubes. *Nanoscale Research Letters*, 10(1), 485.

Wernik, J.M. and Meguid, S.A. (2010). Recent developments in multifunctional nanocomposites using carbon nanotubes. *Applied Mechanics Reviews*, 63(5), 050801.

Xin, F. and Li, L. (2011). Decoration of carbon nanotubes with silver nanoparticles for advanced CNT/polymer nanocomposites. *Composites Part A: Applied Science and Manufacturing*, 42(8), 961–967.

Yazdani, H., Smith, B.E., Hatami, K. (2016). Electrical conductivity and mechanical performance of multiwalled CNT-filled polyvinyl chloride composites subjected to tensile load. *Journal of Applied Polymer Science*, 133(29), 1–12.

Yu, Y., Song, G., Sun, L. (2010). Determinant role of tunneling resistance in electrical conductivity of polymer composites reinforced by well dispersed carbon nanotubes. *Journal of Applied Physics*, 108(8), 084319.

Zeng, X., Xu, X., Shenai, P.M., Kovalev, E., Baudot, C., Mathews, N., Zhao, Y. (2011). Characteristics of the electrical percolation in carbon nanotubes/polymer nanocomposites. *The Journal of Physical Chemistry C*, 115(44), 21685–21690.

Zhao, H. (2015). Preparation and characterization of carbon micro/nano hybrids and their functional composites. PhD Thesis, Université Paris-Saclay, Gif-sur-Yvette.

Zhao, H. and Bai, J. (2015). Highly sensitive piezo-resistive graphite nanoplatelet-carbon nanotube hybrids/polydimethylsilicone composites with improved conductive network construction. *ACS Applied Materials and Interfaces*, 7(18), 9652–9659.

Zhou, J. and Lubineau, G. (2013). Improving electrical conductivity in polycarbonate nanocomposites using highly conductive PEDOT/PSS coated MWCNTs. *ACS Applied Materials & Interfaces*, 5(13), 6189–6200.

8

Electrostrictive Polymer Nanocomposites: Fundamental and Applications

Shenghong YAO and Jinkai YUAN

Centre de Recherche Paul Pascal, CNRS – University of Bordeaux, Pessac, France

8.1. Introduction

All dielectric materials have electroactivity, that is, the ability to changing their dimension or shape in response to an applied electric field (Dang *et al.* 2012). The electroactive polymer (EAP) and its polymer nanocomposites are particularly interesting for applications that range from actuators or sensors to generators, due to their low moduli, high strain capabilities and the ease of processing at low cost and customizable electromechanical coupling properties.

Generally, EAPs have the ability to induce strains that are two orders of magnitude greater than the rigid and fragile electroactive ceramics. They show higher response speed as compared with the shape memory alloys and polymers (Yuan *et al.* 2019). Because of these features, EAPs can be strongly comparable to biological muscles and have been long named "artificial muscles" (Bar-Cohen 2002). A challenge to the community has even been made: to develop a robotic arm actuated by artificial muscles to win an arm wrestling match against a human opponent.

Nanocomposites,
coordinated by Jinbo BAI.

In addition to actuators, EAPs have also revealed their potential in sensory applications such as haptic sensing, blood pressure and pulse rate monitoring and chemical sensing (Wang *et al.* 2016). Furthermore, EAPs even act as key active materials in generators. With ever-increasing functionalities of portable electronics (e.g. wireless sensors and transmitters) and wireless micro-systems, their energy requirements have been dramatically elevated. The use of batteries is troublesome due to the environmental concern and limited lifespan, necessitating their periodic replacement. The obvious solution to this challenge is to develop self-powered systems that completely rely on the energy harvested from the human body or the ambient environment. EAPs have been demonstrated to be able to scavenge vibrational mechanical energy (Lallart *et al.* 2012) and ocean wave energy (Jean *et al.* 2012).

EAPs can be divided into different subgroups based on the crystal class (i.e. centrosymmetric or non-centrosymmetric) that they belong to. As a dielectric material with a center of symmetry is stimulated by an electric field, the symmetry will cancel the movements of cations and anions and does not lead to a net deformation in the crystal. Nevertheless, the chemical bonds are not harmonic and the anharmonicity of the bonds gives rise to second-order effects resulting in a small amount of net deformation of the lattice (Vijaya 2013). The deformation is found to be proportional to the square of the electric field and independent of the direction of the field. This effect is called electrostriction. Because such an anharmonic effect exists in all dielectrics, all dielectrics are electrostrictive materials.

On the other hand, as a non-centrosymmetric dielectric material is subjected to an external electric field, the asymmetric movement of the neighboring ions results in much more significant deformation of the crystal. Such deformation dominates the net deformation induced by the anharmonicity of the chemical bonds (Vijaya 2013). Moreover, it is linearly proportional to the magnitude of the electric field. Reversing the applied field will change the direction of deformation. These materials are called piezoelectrics. A detailed classification of electroactive (dielectric) materials is shown in Figure 8.1.

In this chapter, we focus on electrostrictive polymeric materials. The fundamental issues and promising applications will be addressed. The recent progress toward giant-electrostriction EAPs in the field will be reviewed and some long-standing challenges for their practical applications will be also discussed.

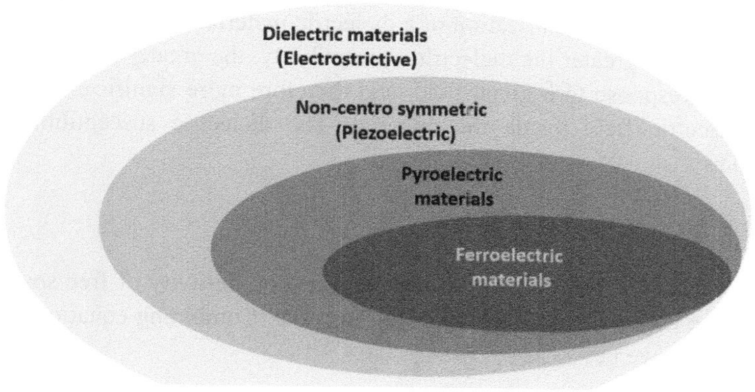

Figure 8.1. *Classification of dielectric materials*

8.2. Electrostriction relations

Electrostriction represents the ability of materials to change their dimensions in response to an applied electric stimulus. As previously mentioned, it is present in all materials, regardless of the symmetry. It is generally defined as a quadratic coupling between strain (S_{ij}) and polarization (P_m) (Damjanovic 1998; Lallart *et al.* 2012).

$$S_{ij} = S^P_{ijkl}T_{kl} + Q_{ijmn}P_mP_n \quad [8.1]$$

$$E_m = \varepsilon'^T_{mn}P_n + 2Q_{klmn}T_{kl}P_n \quad [8.2]$$

where Q_{ijkl} and Q_{klmn} are the polarization-related electrostriction coefficients, S^P_{ijkl} is the elastic compliance, T_{kl} is the stress, E_m is the electric field and ε'^T_{mn} is the inverse of the linear dielectric permittivity.

The polarization P that is induced in a polarizable material by an applied electric field vector E is given by:

$$P_m = \chi_{mk}E_k \quad [8.3]$$

$$P_n = \chi_{nl}E_l \quad [8.4]$$

where χ is the second-rank tensor, known as the dielectric susceptibility of the material. This relation is valid only for linear materials or in a linear limit for nonlinear materials. The dielectric susceptibility χ is a proportionality constant that describes the degree of polarization of a dielectric material in response to an applied electric field. The greater the dielectric susceptibility, the greater the polarization of a material in response to a given field, and therefore more significant reduction of the total electric field inside the material. The dielectric susceptibility can be expressed as

$$\chi = \varepsilon_0(\varepsilon_r - 1) \tag{8.5}$$

where $\varepsilon_0 = 8.854 \times 10^{-12}$ F m^{-1} is the dielectric permittivity of free space and ε_r is the relative dielectric permittivity of the material. Combining equations [8.3] and [8.4] with [8.1], we get

$$S_{ij} = S^E_{ijkl}T_{kl} + M_{ijkl}E_kE_l \tag{8.6}$$

where S^E_{ijkl} is the elastic compliance under constant electric field, M_{ijkl} is the electric field-related electrostriction coefficient and it relates with Q_{ijmn} by

$$M_{ijkl} = \chi_{mk}\chi_{nl}Q_{ijmn} = \varepsilon_0^2(\varepsilon_r - 1)^2 Q_{ijmn} \tag{8.7}$$

On the other hand, the total surface charge density that is induced in the materials by the applied field is given by the dielectric displacement vector D_i,

$$D_i = \varepsilon_0 E_i + P_i \tag{8.8}$$

Combining with equation [8.3], one can deduce

$$\begin{aligned} D_i &= \varepsilon_0 E_i + \chi_{ij}E_j \\ &= \varepsilon_0\delta_{ij}E_j + \chi_{ij}E_j \\ &= (\varepsilon_0\delta_{ij} + \chi_{ij})E_j \\ &= \varepsilon_{ij}E_j \end{aligned} \tag{8.9}$$

where ε_{ij} is the dielectric permittivity of the material and δ_{ij} is Kronecker's symbol.

$$\text{As } i = j,\ \delta_{ij} = 1 \tag{8.10}$$

$$\text{As } i \neq j,\ \delta_{ij} = 0 \tag{8.11}$$

For most of the ferroelectric materials, $\varepsilon_0\delta_{ij} \ll \chi_{ij}$, therefore $\varepsilon_{ij} \approx \chi_{ij}$ (Damjanovic 1998). On the basis of above equations, we can obtain

$$D_i = \varepsilon_{ik}^T E_k + 2M_{ijkl}T_{kl}E_l \quad [8.12]$$

where ε_{ik}^T is the linear dielectric permittivity of materials under constant mechanical stress.

8.3. Determination of the electrostriction coefficient

Experimental determination of the electrostriction coefficient M is typically performed in the actuator mode. The polymer samples are coated with conducting metallic or organic layers on the two opposite surfaces to form a sandwiched capacitor structure. The electrostriction coefficient is therefore characterized by measuring the induced thickness variation under certain magnitudes of high electrical field. In this case, the global strain S is the sum of two distinct contributions, that is, S_E and S_{Max}. The former, S_E, corresponds to the true electrostriction that originates from direct coupling between the polarization and mechanical response, while the latter, S_{Max}, is a result of Maxwell stress that originates from the Coulomb interaction between oppositely-charged compliant electrodes. Therefore, the apparent electrostriction coefficient, M_{app}, is given by

$$M_{app} = S/E^2 \quad [8.13]$$

Plenty of previous research has evaluated the electrostriction coefficients of various EAPs based on the actuator method. For example, Wongtimnoi *et al.* (2011) studied the electrostrictive properties of a polyether-based polyurethane (PU) elastomer and its corresponding composites filled with conductive carbon black (CB) by measuring the thickness strain induced by external electric fields (Figure 8.2). Near the percolation threshold, the composite demonstrated an electrostriction coefficient as high as -1.2×10^{-14} m^2/V^2. Park *et al.* (2008) reported a novel electroactive single-walled carbon nanotube (SWNT)–polyimide composite. This composite intrinsically forms a unimorph during the fabrication process in order to actuate bending deformation in response to an applied electric field. A strain of 2.6% can be obtained at low driving voltages (<1 MV/m). It was claimed that such strong actuation capabilities of the EAPs were gained from the incorporation of SWNTs into these polymers. The electrostriction coefficients (M_{33}) of the EAP composites were estimated from -3.8×10^{-15} to -1.2×10^{-13} m^2/V^2. The high end of these values represent the state of the art.

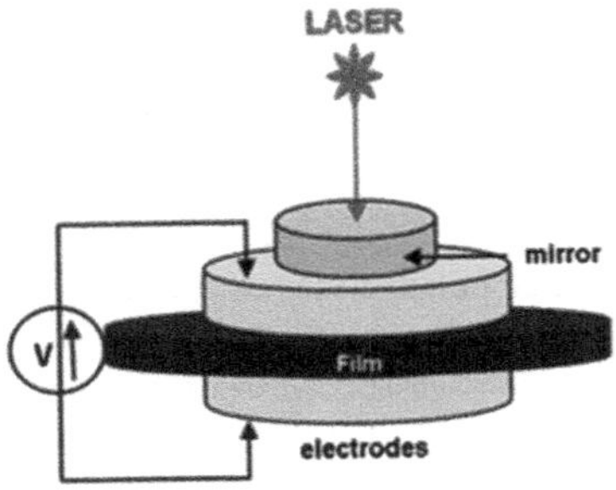

Figure 8.2. *A sandwiched structure used to evaluate the electrostrictive coefficient by measuring the thickness variation in response to the electric field (Wongtimnoi* et al. *2011). © 2011, Elsevier. For a color version of this figure, see www.iste.co.uk/bai/nanocomposites.zip*

The actuator approach can be easily implemented experimentally and is highly efficient to characterize the global electrostriction. Nevertheless, it necessitates the application of a voltage of about several hundred volts. At high voltage, assumptions are needed to determine whether the deformation is dominated by true electrostriction or by the Maxwell stress (Figure 8.3). In practice, it is very difficult to differentiate the true electrostriction from the Maxwell stress experimentally, particularly for heterogeneous dielectric systems. Both of these stresses induce bulk deformations that display a quadratic dependence with the electrical field.

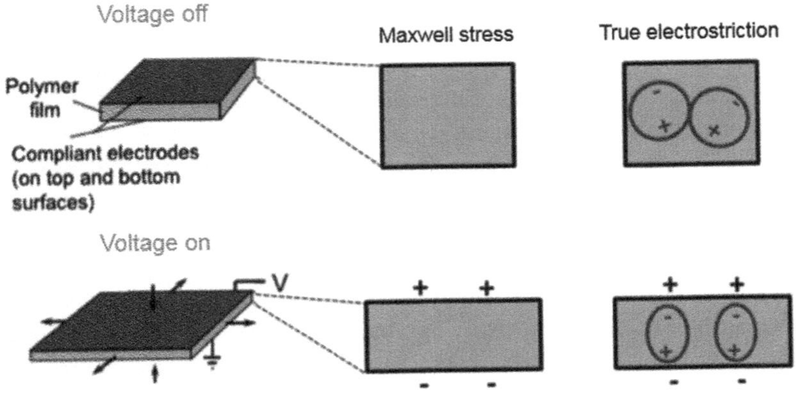

Figure 8.3. *An illustration demonstrating the different mechanisms for the actuation of the materials in response to an applied electric field. For a color version of this figure, see www.iste.co.uk/bai/nanocomposites.zip*

According to the literature, the actuation for elastomeric materials with low permittivity ($\varepsilon_r' < 3$), such as silicones, natural rubber latex and acrylates, results mainly from Maxwell stress. On the other hand, for high permittivity ($\varepsilon_r' > 5$) polar polymers, such as polyvinylidene fluoride (PVDF) and PU, the actuation is dominated by the electrostriction mechanism and Maxwell stress as well.

The electrostriction also describes the ability of materials to change their dielectric properties with deformation applied upon them, as shown in equation [8.12]. By replacing the D_i with equation [8.9], one can obtain

$$\varepsilon_{ij}E_j = \varepsilon_{ij}^T E_j + 2M_{ij}T_{ij}E_j \qquad [8.14]$$

$$\Delta\varepsilon_{ij} = \varepsilon_{ij} - \varepsilon_{ij}^T = 2M_{ij}T_{ij} \qquad [8.15]$$

$$M_{ij} \approx \frac{\Delta\varepsilon_{ij}}{2T_{ij}} \approx \Delta\varepsilon_{ij}/2Y\varphi_j \qquad [8.16]$$

where Y is the Young's modulus and φ_j is the mechanical strain. In contrast to the actuator mode widely used in previous studies, we have developed a new method based on equation [8.16], that is, measuring the permittivity variation with applied mechanical deformation. This method allows for direct measurements of the true electrostriction coefficients with a negligible contribution of the Maxwell stress. In our measurements, a voltage of only 1 V is applied, which corresponds to an electrical field below 10^{-3} V/μm. The significance of Maxwell stress can be estimated by comparing the strain resultant from Maxwell stress and the strain mechanically applied by the tensile test mechanical instrument. The strain S_{Max} resulting from Maxwell stress is given by

$$S_{Max} = \frac{\varepsilon_{ij}E^2}{Y} \qquad [8.17]$$

where E and Y, respectively, are the electrical field and the Young's modulus of the material. This strain (~10^{-6}%) is totally negligible compared to the strain mechanically applied (~10%). In the case of an isotropic electrostrictive polymer film subjected to a stress τ_{11} and to an electrical field along its thickness direction E_3 (Figure 8.4(a)), the electrostriction coefficient can be expressed as

$$M_{31} = \Delta\varepsilon_{31}/2Y\varphi_{11} \qquad [8.18]$$

While in an isotropic electrostrictive polymer film subjected to a stress τ_{33} and to an electrical field along its thickness direction E_3 (Figure 8.4(b)), the electrostriction is caculated as

$$M_{33} = \Delta\varepsilon_{33}/2Y\varphi_{11} \qquad [8.19]$$

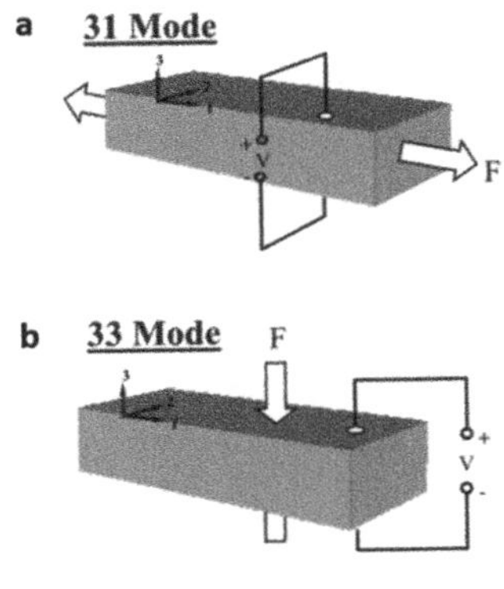

Figure 8.4. *Two modes used to characterize the electrostriction coefficient by measuring the permittivity variation with mechanical deformation. For a color version of this figure, see www.iste.co.uk/bai/nanocomposites.zip*

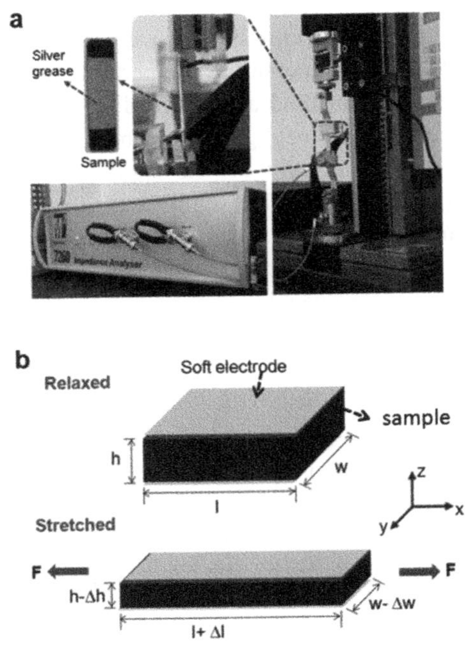

Figure 8.5. *a) Setups for the measurement of permittivity variation during mechanical stretching. b) The mechanism by which a parallel plate capacitor changes its capacitance in response to a mechanical stretching (Yuan* et al. *2018). © 2018, American Chemical Society. For a color version of this figure, see www.iste.co.uk/bai/nanocomposites.zip*

So the question arises: How can we measure the variation of dielectric permittivity? We have combined dielectric spectroscopy with the tensile test to meet this end (Figure 8.5(a)) (Yuan *et al.* 2018). The change in capacitance of a parallel plate capacitor relies on Poisson contraction as well as permittivity variation (Figure 8.5(b)). Longitudinal stretching of the capacitor causes elongation in length yet shrinkage in width and thickness due to Poisson's effect.

The linear mechanics of the mechanical stretching process can be described as

$$\varphi_x = \frac{\Delta l}{l} \quad [8.20]$$

$$\varphi_y = \varphi_z = \nu\varphi_x \quad [8.21]$$

$$l_{stretch} = l + \Delta l = l + \varphi_x l \quad [8.22]$$

$$w_{stretch} = w - \Delta w = w - \nu\varphi_x w \quad [8.23]$$

$$h_{stretch} = h - \Delta h = h - \nu\varphi_x h \quad [8.24]$$

where ν is Poisson's ratio; φ_x is the strain; and l, w and h are the dimensions of the capacitor in a relaxed state. Considering the dimensions of the capacitor in its stretched state, one can deduce the value of the capacitance C_{stretch}:

$$C_{stretch} = \varepsilon_0\varepsilon_{stretch}\left(\frac{w_{stretch}\, l_{stretch}}{h_{stretch}}\right)$$

$$= \varepsilon_0\varepsilon_{stretch}\left(\frac{(w - \nu\varphi_x w)(l + \varphi_x l)}{h - \nu\varphi_x h}\right)$$

$$= \varepsilon_0\varepsilon_{stretch}\frac{wl}{h}(1 + \varphi_x) \quad [8.25]$$

where ε_0 is the permittivity of free space and $\varepsilon_{stretch}$ is the relative permittivity in the stretched state. This electromechanical equation directly relates the capacitance to the permittivity, applied strain and initial geometry of the capacitor. Despite the fact that this equation is derived from Poisson contraction, it does not depend on the value of Poisson's ratio in the case of isotropic dielectric materials. Therefore, by measuring the capacitance at different strains, we can directly determine the permittivity variations ($\Delta\varepsilon$) of the material in response to mechanical deformations. This allows for calculation of true electrostrictive coefficients of a material directly through the aforementioned equation [8.18] or [8.19].

8.4. The route to high electrostriction materials

Electrostrictive materials have been widely used to efficiently convert electrical and mechanical energy (Zhang *et al.* 2002). As previously mentioned, all dielectric materials exhibit electrostriction. However, certain materials, known as relaxor ferroelectrics, have a high electrostriction coefficient (Zhang *et al.* 1998). Today, the most used electrostrictive ceramics are lead-based materials such as lead magnesium niobate (PMN), lead magnesium niobate-lead titanate (PMN-PT) and lead lanthanum zirconate titanate (PLZT). In spite of the high electrostrictive performances, they exhibit several drawbacks, such as high sintering temperaure, brittleness and lead-related environmental concerns. Lead-free and flexible electrostrictive materials are always in high demand, particularly for large-stroke actuation applications. In this regard, the polymeric materials, such as dielectric elastomers (Pelrine *et al.* 2000; Lehmann *et al.* 2001; Brochu and Pei 2010; Luna *et al.* 2015, 2017; Yuan *et al.* 2015), high-k elastomer composites (Clarke *et al.* 2001; Huang *et al.* 2009; Kohlmeyer *et al.* 2009; Molberg *et al.* 2010; Wongtimnoi *et al.* 2011; Kuang *et al.* 2015; Romasanta *et al.* 2015) and PVDF and its copolymers (Zhang *et al.* 1998; Li 2003; Huang *et al.* 2004; Minh Quyen *et al.* 2015), are more promising. However, most of the current electrostrictive polymers suffer from a low electromechanical coupling coefficient ($M \sim 10^{-18}$ m^2/V^2 at 100 Hz) (Zhang *et al.* 1997; Guillot and Balizer, 2003; Huang *et al.* 2004) due to their weak intrinsic polarization response, which severely restricts their applicability for practical devices. A major challenge facing the community is the development of electromechanically-active polymeric materials with giant electrostriction (Yuan *et al.* 2018).

As discussed in the previous section, the electrostriction is a result of the change of the polarization in materials in response to the external mechanical or electrical stimuli. The electromechanical properties can thus be optimized by improving the polarization response in polymers through designing structures on different scale levels.

NASA has developed a new electrostrictive graft elastomer (Su *et al.* 2000, 2002) which is approximately 30 times stiffer than classic PU. These strong elastomers have potential employment as an actuator that provides much higher mechanical energy and power output than soft neat elastomers. As illustrated in Figure 8.6, the graft elastomer contains flexible backbone chains and grafts that are the side chains of each backbone. The grafts physically cross-link with each other and eventually form crystal units. Both the backbone chain and the crystal graft units have polarized monomers. These monomers contain atoms with electric partial charges, giving rise to dipole moments. These dipole moments will rotate to align with the external electric field and therefore stimulate backbone chain reorientation and induce a much larger electrostrictive strain than neat PU (Wang *et al.* 2004).

Another example could be the dielectric elastomer gel. Kim *et al.* (2011) introduced periodic nanoscale interfaces to dielectric elastomer gels and achieved an unexpected high electrostriction coefficient by creating numerous interfaces between dielectric mismatched nanostructured domains, resulting in a large electromechanical strain response at a relatively low electric field. A TEM picture showed the nanostructured domains in a thermoplastic elastomer gel composed of 20 wt% poly(styrene-block-ethylbutylene-block-styrene) (SEBS) and 80 wt% white mineral oil. The strong electrostrictive effect is attributed to the inhomogeneous electric field in the film thickness direction because of the high density of interfaces between dielectric mismatched periodic nanoscale domains. The demonstrated electrostrictive coefficient can be two orders of magnitude higher than that of neat PU.

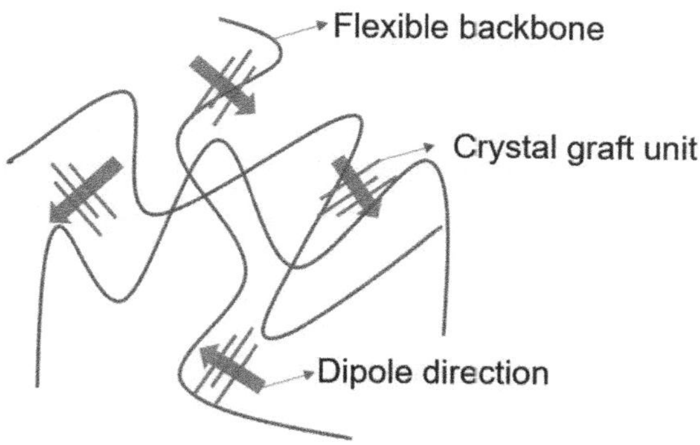

Figure 8.6. *Illustrative structure of the electrostrictive graft elastomer (Wang* et al. *2004). For a color version of this figure, see www.iste.co.uk/bai/nanocomposites.zip*

Normal ferroelectric poly(vinylidene fluoride-trifluoroethylene) P(VDF-TrFE) polymers have excellent piezoelectric properties. However, after the introduction of non-polar molecular structure in the ferroelectric phase (Zhang *et al.* 1998) or the chemical copolymerization of a third bulky monomer (Xia *et al.* 2002), it can be converted to relaxor ferroelectric polymers with large electrostriction. The improved electrostriction is due to the difference in the lattice constant between the polar and non-polar phases and to the gradual increase in field-induced polarization in the relaxor P(VDF-TrFE) polymer.

Though improvements of the electrostriction effect have been achieved based on changing the polymer composition, a versatile strategy actually relies on the composite approach, that is, mixing electroactive nanofillers with polymer matrices. It allows for a large tunability by varying the identity, shape and content of nanoadditives.

Giant permittivity can actually be obtained in electrically heterogeneous systems composed of conducting particles and insulating polymer matrices (Yuan *et al.* 2011, 2014, 2017; Luna *et al.* 2015, 2017) The permittivity of a composite material diverges at the percolation threshold (Dang *et al.* 2012, 2013; Luna *et al.* 2015, 2017; Yuan *et al.* 2015, 2017, 2018; Yuan 2017). Systems become critically sensitive at this point, thereby providing the possibility for giant electrostriction (Luna *et al.* 2017).

A large amount of research have been conducted by randomly mixing conducting particles, such as metal nanoparticles and carbon-based nanomaterials, with polymers to reach a high permittivity. The improved electromechanical performance can be associated with the variation of the local electrical field as well as the strong interfacial polarization as a result of charge accumulation (Lallart *et al.* 2012). Typical solution methods were used to randomly mix the nanoparticles with polymer matrices. Unfortunately, the electrostriction coefficient experimentally achieved until now remains low. This is likely due to a lack of optimization of the morphology of the near-percolated networks formed by the particles.

Recently, we reported a soft electrostrictive elastomer nanocomposite composed of reduced graphene oxide (rGO) in a polydimethylsiloxane (PDMS) matrix. The advantage of using graphene is based on the essential feature of its large surface area, which is critical to achieve large polarization in composites. A diblock copolymer aminopropylmethylsiloxane-*b*-dimethylsiloxane (PAPMS-*b*-PDMS) was chosen to adsorb onto graphene oxide (GO) nanosheets and stabilize their dispersion in composite precursor solutions. Such interaction is based on the electrostatic attraction between negative charges of GO nanosheets and positive charges of the copolymer (Figure 8.7(a)). As such, the GO nanosheets continuously transferred from water phase into diethyl ether phase as this interaction at the water/oil interface proceeded (Figure 8.7(b)) (Yuan *et al.* 2015).

The obtained GO suspension is well suited to preparing PDMS matrix composites without showing any reaggregation of GO nanosheets during the process. Moreover, with increasing concentration of such a GO suspension, a nematic liquid crystal (LC) phase can form even with a low GO content (Figures 8.7(c) and (d)).

Experimentally, we assessed the percolation transition in rGO/PDMS composites. The margin improvement of conductivity at 1.0 wt% indicates that this loading is well below the percolation threshold. However, at this rGO-PDMS mass ratio, LC transition has already taken place (Figure 8.7(d)), indicating that the flakes with large anisotropy tend to self-assemble into nematic LCs before they form a percolated path along the direction of the conductivity measurement. This phenomenon would fundamentally change our views on the interest of graphene for conductive composites. Such competition can, nevertheless, be positively exploited to develop dielectric composites. The ordered liquid crystalline graphene structure is expected to promote the formation of effective microcapacitors to afford a very high permittivity (~2000 at 0.1 Hz) (Yuan *et al.* 2015).

It is found that graphene-doped PDMS exhibits very large variations of permittivity within a very small strain range due to the liquid crystalline alignment of the graphene sheets within the polymer. Such large variations yield an ultralarge electrostriction coefficient ($M \sim 10^{-14}$ m^2/V^2 at 0.1 Hz), superior to those (from 10^{-18} m^2/V^2 to 10^{-15} m^2/V^2 at 0.1 Hz) obtained in conventional polymer nanocomposites (Guiffard *et al.* 2006; Park *et al.* 2008; Wongtimnoi *et al.* 2011, 2013; Guyomar *et al.* 2012; Lallart *et al.* 2012; Jaaoh *et al.* 2015, 2016).

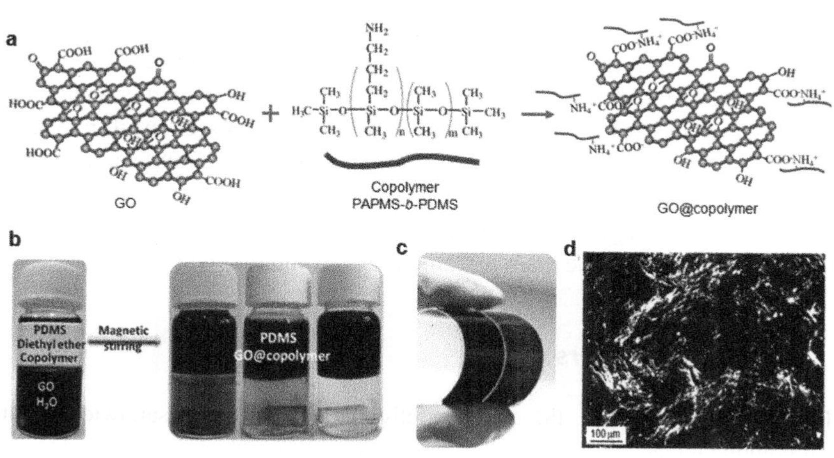

Figure 8.7. *a) Schematic illustration of the electrostatic attraction between negative charges of GO nanosheets and positive ones of the PAPMS-b-PDMS copolymer. b) Photograph shows the phase transfer of GO flakes from the aqueous to the organic phase. c) Photograph of soft PDMS composites with GO (brown) and rGO (black) nanosheets at a loading of 1 wt%. d) Optical micrographs between crossed polarizers of GO suspensions in PDMS monomer at a concentration of 1 wt% (Yuan* et al. *2015). For a color version of this figure, see www.iste.co.uk/bai/nanocomposites.zip*

8.5. Applications of electrostrictive materials

Electrostrictive polymeric materials have been well-known for their significant advantages of facile proccessibility, short response times (less than milliseconds), low hysteresis and the ability to generate high strain levels (>7%), which is far above those of traditional inorganic electrostrictive materials or piezoceramics. Therefore, ever-increasing research attention has been paid to the community of electrostrictive polymers during recent decades because of their promising applications, ranging from functional actuators and capacitive sensors to integrated microelectromechanical systems. Recently, electrostrictive polymers have even been shown to be potentially useful for harvesting energy from vibration sources to power electronic devices.

8.5.1. *Actuators*

Dielectric elastomer, as one type of EAP, has been widely used to develop fuctional actuators including artifical muscles. Dielectric elastomer deforms due to the electrostatic interaction between two electrodes with opposite electric charges. In earlier years, most of the focus was on actuator configurations. Electroded dielectric elastomer films can be rolled into a scroll, stretched over a frame, or laminated on a flexible substrate to form unimorphs and bimorphs or even formed into a tubular shape. In more recent years, the focus has turned to using the material properties, particularly the electrostriction coefficient (see the previous section) to improve the actuating performances. In a review paper, O'Halloran *et al.* (2008) summarized the operating principle and actuation mechanisms behind the actuator technology, highlighted its advantages over other existing actuator technologies, identified the challenges and examined the main research focus within the field.

8.5.2. *Capacitive sensors*

In addition to actuators, the electrostrictive polymers can be sandwiched with a pair of compliant electrodes to create a variable capacitor, which can be used to detect pressure or deformation in the change of capacitance. Such pressure sensors strategically placed on a flexible synthetic skin working in the pressure range of 1 kPa to 1 MPa would be able to detect fingertip texture, hand grip and in-shoe pressures.

According to equation [1.25], for a sensor fabricated in three layers with a dielectric between the electrodes, the capacitance will be proportional to strain under a simple stretching. As we know, the small strain sensing can be also realized by electrical resistance measurement of elastomer composites embedded with conducting nanoparticles. However, for large strain sensing that involves filled elastomers, the measurement will be nonlinear due to geometric effect and also transient changes in resistivity of the electrodes.

Capacitance measurement offers an opportunity for reliable measurement that is nearly unaffected by transient changes in electrode resistance. The capacitive strain gauge factor κ can be written as

$$\kappa = \frac{\Delta C / C_0}{\varphi_x}$$

$$= \frac{\Delta\varepsilon/\varepsilon}{\varphi_x} + \frac{\Delta\varepsilon}{\varepsilon} + 1 \quad [8.26]$$

where $\Delta\varepsilon$ is the permittivity variation. For non-polar dielectrics like silicon rubber, $\Delta\varepsilon = 0$, so the $\kappa = 1$. While for electrically heterogeneous dielectrics, $\Delta\varepsilon \neq 0$, the capacitance variation ΔC is determined by the geometry effect coupled with the true electrostriction effect of the dielectrics, as the case of near-percolated elastomer composites with conducting nanoparticles. Plenty of reported capacitive sensors are based on pure silicons sandwiched by diverse compliant electrodes. For instance, Bao's group reported a capacitive sensor with transparant SWNT films as compliant electrodes. This sensor can detect a pressure up to 50 kPa and strain up to 50%. The capacitance linearly increases with the strain, showing a slope near 1. In spite of the fact that is not very sensitive to extremely low pressures, it is still able to detect pressures generated by a fly (20 mg). The transparency greatly expands the scope of applications, such as touch screens and electronic skins with various options of skin color.

In order to improve the sensitivity of capacitive sensors, the key is to employ soft materials with a high electrostriction coefficient. According to equation [8.26], combining the geometry effect and electrostriction to induce a large capacitance variation for a given strain level is promising. Based on the near-percolated rGO composite demonstrated in Figure 8.7, we showed a highly efficient capacitive sensor in response to a cyclic mechanical tensile loading (Figure 8.8(a)). Such a sensor device was stretched to a strain of 4%, after which the deformation was released with the same rate as stretching. Such a cycle was repeated 20 times. It was noticed that during the first four cycles, the capacitance variation slightly decreased from 2.4 to 2.2 nF yet hereafter tended to remain stable (Figure 8.8(b)). Recalling that the capacitance at relax state is 4.1 nF, a constant capacitive gauge factor of –13.6 was obtained. Compared with the capacitive sensors based only on the geometry effect (gauge factor <1), the electrostrictive composite sensors show a much higher sensitivity, which derives from their ultrahigh electrostriction.

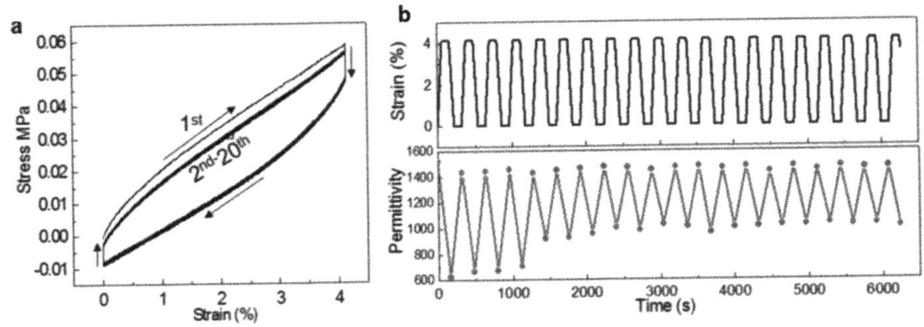

Figure 8.8. *a) Stress as a function of strain under a cyclic tensile loading-deloading. b) Time dependence of the strain and relative permittivity at 0.1 Hz over the 20 stretching/releasing cycles (Yuan* et al. *2018). © 2018, American Chemical Society. For a color version of this figure, see www.iste.co.uk/bai/nanocomposites.zip*

8.5.3. *Mechanical energy harvesting*

Dielectric elastomers have demonstrated great promise as actuator and sensory materials. Nevertheless, their potential in converting mechanical enegy into electricity in a generator mode has been used less. In practice, dielectric elastomer acts electrically as variable capacitors do. If a low voltage is placed on a pre-strained dielectric elastomer, the contraction works against the electrostatic field stress and raises the voltage of the charge, generating electrical energy. Such simplicity, combined with demonstrated high energy density and high efficiency (Pelrine *et al.* 2001; Lallart *et al.* 2012) suggests that dielectric elastomers are promising for a wide range of energy-harvesting applications. Compared with competing energy-harvesting technologies, such as electromagnetic or piezoelectric ones, dielectric elastomers show less stiffness, meaning they can be applied to large deformations and operate even at frequencies as low as 0.1 Hz (ocean waves). Their energy density can reach 0.4 J/g, which is one order of magnitude higher than those of the counterparts. Meanwhile their energy conversion efficiency can be higher than 50%, while that of an electromagnetic technology is usually less than 20%.

Figure 8.9 shows the performance limits and also the basic operational cycles of a dielectric elastomer generator. In detail, there are four basic steps in the simplest operational cycles: (1) the film is stretched to its maximum stretch state; (2) a voltage or charge is applied to the film; (3) the film relaxes its internal elastic energy with constant charge, constant voltage or constant field; and (4) the charge is removed from the film to return it to its initial state. The horizontal axis of

Figure 8.9 is a variable related to the change in capacitance, while the vertical axis is the square of the voltage or electric field across the film. Therefore, the shaded blue area represents the amount of extracted energy. The net amount of extracted energy per unit volume of film for the constant charge (u_Q), constant voltage (u_V), and constant field (u_E) cycles, respectively, are (Kornbluh *et al.* 2011, 2012) given as

$$\mu_Q = \frac{1}{2}\varepsilon E_{max}^2\left[\frac{\gamma^2-1}{\gamma^2}\right] \quad [8.27]$$

$$\mu_V = \frac{1}{2}\varepsilon E_{max}^2\left[\frac{\gamma^2-1}{\gamma^2}\right] \quad [8.28]$$

$$\mu_E = \varepsilon E_{max}^2 \ln(\gamma) \quad [8.29]$$

where E_{max} and ε are, respectively, the maximum electric field and permittivity of the film samples. γ is the ratio of stretched area/unstretched area.

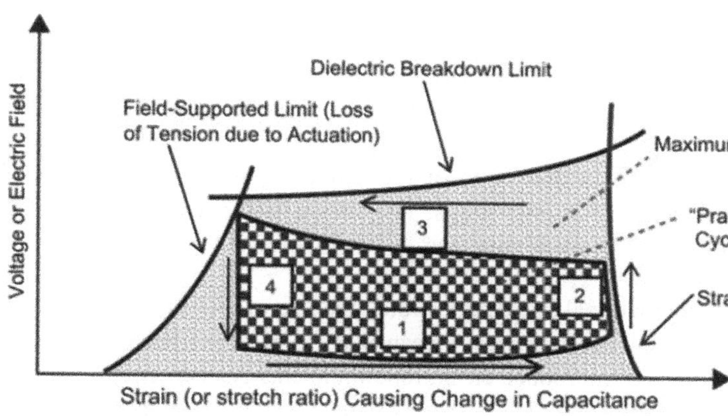

Figure 8.9. *The performance limits and operational cycles of a dielectric elastomer generator (Kornbluh* et al. *2012). For a color version of this figure, see www.iste.co.uk/bai/nanocomposites.zip*

On the basis of previous discussion, desirable dielectric elastomers for mechanical energy harvesting should have high breakdown strength, high permittivity and low dielectric losses. Meanwhile, they should sustain high flexibility so that they can be deformed with small forces, low viscoelastic losses and low creep and stress relaxation effect. In recent decades, the most common candidate materials considered for generator applications are the ones based on commercial formulations of acrylics and silicones (Carpi *et al.* 2008) as well as PU-based polymers (Wongtimnoi *et al.* 2011, 2013).

Pelrine *et al.* (2001) demonstrated the fundamentals and applications of neat silicone as a generator. They used the constant charge cycle to harvest the stored mechanical energy as the prestrained silicone elastomer relaxed. The energy harvesting circuit is shown in Pelrine *et al.* (2001). The film was initially biased with a bias voltage supplied through a diode. A drain resistor in parallel exhibits a huge resistance so that the charge that is drained during contraction can be negligible. The dielectric elastomers have a low enough leakage current so that the decay back to the bias voltage can take tens of seconds. The extracted energy can be easily manifested by the boosting and decay of the voltage on the drain resistor.

Dielectric elastomers are well suited to applications in which electrical power comes from relatively large motions, such as those produced by wind energy, waves and human activities. Kornbluh *et al.* (2011, 2012) have developed a heel-strike generator using electrode-coated silicone polymers. Such a generator can be installed in a normal shoe or boot (Ashley 2003).

The compression of the heel during normal walking can harvest power from human activity yet does not add any physical burden to the wearer. The device produced an electrical output of 0.8 J per step (the power is 1 W). The dielectric elastomer generator is an array of multilayer diaphragms that transfer the compression of the heel into deflection of the active layers, which consist of 20 stacked layers of dielectric elastomer films. The device used prestrained VHB 4910 acrylic and performed with a maximum energy density of about 0.3 J/g.

Recently, SRI International has demonstrated the use of dielectric elastomer for harvesting the energy of ocean waves. Ocean wave power has been called "blue energy" and has potential to produce clean, renewable energy in an environmentally sound manner. Harvesting such blue energy is critical as the world moves toward a more secure and sustainable energy future. It was shown that the ocean wave energy could potentially meet 10% of the world's total electricity demand. The U.S. Electric Power Research Institute (EPRI) estimated the viable resource in the United States at 255 TWh per year (6% of demand), comparable to the energy currently generated in the United States by conventional hydropower (Hagerman and Bedard 2003). Based on the direct conversion of ocean wave to mechanical stretching and contract of the dielectric elastomer, Kornbluh *et al.* have developed a proof-of-principle generator system (Kornbluh *et al.* 2011). This system is shown in Figure 8.10. The system was tested at sea in the Pacific Ocean near Santa Cruz, California. It produced an output of >25 J by using about 220 g of active dielectric elastomer material, giving rise to an energy density of more than 0.1 J/g. The efficiency of the energy-harvesting circuit used was 78%. The overall performance suggests that dielectric elastomers may indeed be practical for large-scale power generation.

Figure 8.10. *Ocean wave power generator based on electrode coated dielectric elastomers (Kornbluh* et al. *2011). For a color version of this figure, see www.iste.co.uk/bai/nanocomposites.zip*

By recognizing that high permittivity is desirable for mechanical energy harvesting, many researchers have focused on improving the energy conversion capabilities of elastomers upon incorporating diverse functional nanoparticles, such as CB (Wongtimnoi 2011), barium titanate (Yin *et al.* 2017; Yang *et al.* 2019) and rGO (Nesser *et al.* 2018).

Yin *et al.* (2017) targeted generators with improved performance of mechanoelectric energy conversion by designing a ternary composite, which was composed of thermoplastic PU, barium titanate (BT) nanoparticles and dibutyl phthalate (DBP) plasticizers. They simultaneously achieved a 28% increase in permittivity and a 43% decrease in Young's modulus by blending the PU with 25 phr of BT and 60 phr of DBP. Both of them are very desirable for improving the mechanical energy-harvesting capability. By simply stretching the sample to an area expansion of 80% and appling a bias voltage of 900 V (the circuit is shown in Figure 8.11(a)), the terpolymer with 15% of BT and 60% of DBP exhibited a harvested energy density up to 1.71 mJ/cm^3 and an efficiency of 0.45%. The harvested energy can light an LED array, consisting of 30 blue LEDs in series (Figure 8.11(b)). As a comparision, they tested the widely investigated dielectric elastomer, VHB4910, under the same stretching-releasing operations and only obtained an energy density of 0.21 mJ/cm^3. These results clearly validate a principle that composite elastomers with tailored properties are indeed desirable for developing advanced generators.

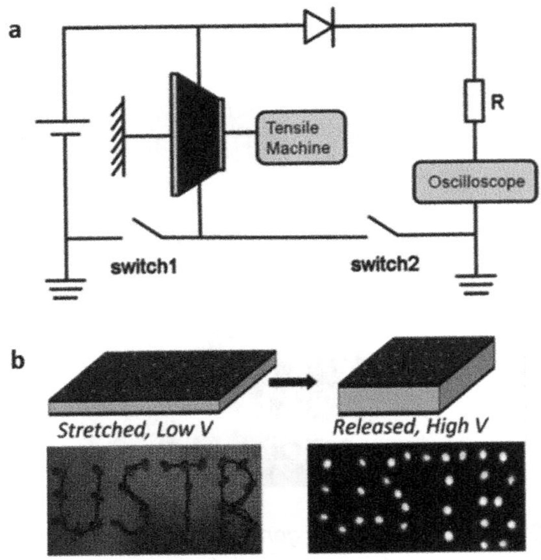

Figure 8.11. *a) The electric circuit used for mechanical energy harvesting. b) 30 blue LEDs in series can be lit by the harvested energy during each cycle (Yin* et al. *2017). © 2017, American Chemical Society. For a color version of this figure, see www.iste.co.uk/bai/nanocomposites.zip*

In addition to the high-*k* nanoparticles, conductive nanofiller-based near-percolative composites have also shown improved electrostrictive properties. They are potentially promising that they can be used for developing generators. Very recently, Nesser *et al.* (2018) integrated electrostrictive nanocomposites, composed of rGO dispersed in polydimethylsiloxane, into all-organic microcantilever resonators specifically designed for mechanical energy harvesting from ambient vibrations. Figures 8.12(a) and (b) present the top and side views of the micro-cantilever, consisting of electrostrictive PDMS/rGO composites coated by two layers of conducting polymer electrodes. The organic micro-electro-mechanical systems (MEMSs) were fabricated at low cost and in an all-organic approach by means of printing methods associated with xurography. The low resonant frequency (<100 Hz) of the resulting microcantilevers can be realized by simply incorporating a seismic mass at their tip. With the help of the constant voltage circuit (Figure 8.12(c)), the MEMS contilever provides a power of more than 6 $\mu W/cm^3$ for an acceleration amplitude of 1 g at the MEMS resonant frequency of 17 Hz (Figure 8.12(d)). The demonstrated output power density is higher than that extracted from PU-based electrostrictive materials (Jaaoh *et al.* 2016), and is on the same level as that generated by PVDF terpolymers (Cottinet *et al.* 2011).

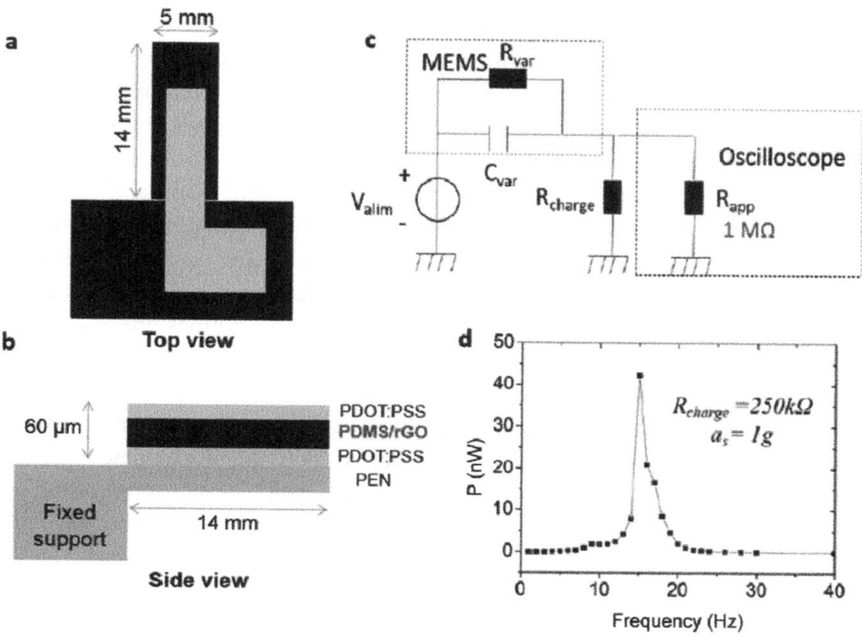

Figure 8.12. *a) Top and b) side view of the micro-electro-mechanical systems (MEMS) cantilever, consisting of electrostrictive PDMS/rGO composites coated by two layers of electrodes. For a color version of this figure, see www.iste.co.uk/bai/nanocomposites.zip*

In spite of the tremendous progress of the generators based on the electrostrictive polymeric materials, there are still several challenges to be addressed in the future. Lifetime and reliability issues may be the greatest challenge to adoption of dielectric elastomers. Tradeoff performances could be made by extending the lifetimes yet this has seldom been characterized in the literature. Reduction of the dielectric losses of electrostrictive materials can be another key issue for improving the output and efficiency of generators. Another challenge relies on the design of energy harvesting circuits. In many cases, tradeoffs will be necessary between circuit complexity and cost. Usually, it has been preferred that the former has high efficiency yet this increases the cost of the generators. Furthermore, the energy-harvesting cycles should be optimized according to the properties of materials and the environmental mechanical resources.

8.6. Conclusions and perspectives

All dielectric materials have the electrostriction effect. This chapter starts from the classification of dielectric mateials and then reviews the fundamental and application-related issues for electrostrictive materials. Particular attention is paid to the means of characterizing the electrostriction coefficient as well as the routes to high electrostriction materials. Compared to changing the polymer composition, the composite approach can be a more versatile strategy. It allows for a large tunability of the electrostriction of polymers by varying the identity, shape and content of nanoadditives. At the end, actuation, sensory and energy-harvesting applications of electrostrictive materials are summerized and examples for each application are given. An emphasis was given to the generator application. Challenges for further improving the generator performances and large-scale applications are discussed in detail. From a more general point of view, we note that the fabrication of nanocomposites is usally guided by intuitive expectations and empirical approaches. Advanced modeling should be further developed in order to provide a more rational guidance of experimental efforts. It is expected that the development of new formulation and microfabrication technologies will help in the optimization and control of nanocomposite structures to achieve enhanced electrostrictive properties in the years to come.

8.7. References

Ashley, S. (2003). Artificial muscles. *Scientific American*, 289(4), 52–59.

Bar-Cohen, Y. (2002). Electroactive polymers as artificial muscles: A review. *Journal of Spacecraft and Rockets*, 39(6), 822–827.

Brochu, P. and Pei, Q. (2010). Advances in dielectric elastomers for actuators and artificial muscles. *Macromolecular Rapid Communications*, 31(1), 10–36.

Carpi, F., DeRossi, D., Kornbluh, R.D., Pelrine, R., Sommer-Larsen, P. (2008). *Dielectric Elastomers as Electromechanical Transducers. Fundamentals, Materials, Devices, Models and Applications of an Emerging Electroactive Polymer Technology*. Elsevier Press, Amsterdam.

Clarke, J., Clarke, B., Freakley, P.K., Sutherland, I. (2001). Compatibilising effect of carbon black on morphology of NR-NBR blends. *Plastics, Rubber and Composites*, 30(1), 39–44.

Cottinet, P.-J., Lallart, M., Guyomar, D., Guiffard, B., Lebrun, L., Sebald, G., Putson, C. (2011). Analysis of AC-DC conversion for energy harvesting using an electrostrictive polymer P(VDF-TrFE-CFE). *IEEE Transactions on Ultrasonics, Ferroelectrics and Frequency Control*, 58(1), 30–42.

Damjanovic, D. (1998). Ferroelectric, dielectric and piezoelectric properties of ferroelectric thin films and ceramics. *Reports on Progress in Physics*, 61(9), 1267–1324.

Dang, Z., Yuan, J., Zha, J., Zhou, T., Li, S., Hu, G. (2012). Fundamentals, processes and applications of high-permittivity polymer matrix composites. *Progress in Materials Science*, 57(4), 660–723.

Dang, Z., Yuan, J., Yao, S., Liao, R. (2013). Flexible nanodielectrics materials with high permittivity for power energy storage. *Advanced Materials*, 25, 6334–6365.

Guiffard, B., Seveyrat, L., Sebald, G., Guyomar, D. (2006). Enhanced electric field-induced strain in non-percolative carbon nanopowder/polyurethane composites. *Journal of Physics D – Applied Physics*, 39(14), 3053–3057.

Guillot, F.M. and Balizer, E. (2003). Electrostrictive effect in polyurethanes. *Journal of Applied Polymer Science*, 89(2), 399–404.

Guyomar, D., Cottinet, P.J., Lebrun, L., Putson, C., Yuse, K., Kanda, M., Nishi, Y. (2012). The compressive electrical field electrostrictive coefficient M33 of electroactive polymer composites and its saturation versus electrical field, polymer thickness, frequency, and fillers. *Polymers for Advanced Technologies*, 23(6), 946–950.

Hagerman, G. and Bedard, R. (2003). Guidelines for preliminary estimation of power production, by offshore wave energy conversion devices. EPRI Report 297213, Electric Power Research Institute.

Huang, C., Klein, R., Xia, F., Li, H.F., Zhang, Q.M., Bauer, F., Cheng, Z.Y. (2004). Poly(vinylidene fluoride-trifluoroethylene) based high performance electroactive polymers. *IEEE Transactions on Dielectrics and Electrical Insulation*, 11(2), 299–311.

Huang, X., Xie, L., Jiang, P., Wang, G., Liu, F. (2009). Electrical, thermophysical and micromechanical properties of ethylene-vinyl acetate elastomer composites with surface modified BaTiO3 nanoparticles. *Journal of Physics D – Applied Physics*, 42(24), 245407.

Jaaoh, D., Putson, C., Muensit, N. (2015). Deformation on segment-structure of electrostrictive polyurethane/polyaniline blends. *Polymer*, 61, 123–130.

Jaaoh, D., Putson, C., Muensit, N. (2016). Enhanced strain response and energy harvesting capabilities of electrostrictive polyurethane composites filled with conducting polyaniline. *Composites Science and Technology*, 122, 97–103.

Jean, P., Wattez, A., Ardoise, G., Melis, C., Van Kessel, R., Fourmon, A., Barrabino, E., Heemskerk, J., Queau, J.P. (2012). Standing wave tube electro active polymer wave energy converter. *Conference on Electroactive Polymer Actuators and Devices (EAPAD)*, San Diego, USA, March 12–15.

Kim, B., Park, Y.D., Min, K., Lee, J.H., Hwang, S.S., Hong, S.M., Kim, B.H., Kim, S.O., Koo, C.M. (2011). Electric actuation of nanostructured thermoplastic elastomer gels with ultralarge electrostriction coefficients. *Advanced Functional Materials*, 21(17), 3242–3249.

Kohlmeyer, R.R., Javadi, A., Pradhan, B., Pilla, S., Setyowati, K., Chen, J., Gong, S. (2009). Electrical and dielectric properties of hydroxylated carbon nanotube-elastomer composites. *Journal of Physical Chemistry C*, 113(41), 17626–17629.

Kornbluh, R.D., Pelrine, R., Pei, Q.B., Heydt, R., Stanford, S., Oh, S.J., Eckerle, J. (2002). Electroelastomers: Applications of dielectric elastomer transducers for actuation, generation and smart structures. *Smart Structures and Materials 2002 Conference*, San Diego, USA, March 18–21.

Kornbluh, R.D., Pelrine, R., Prahlad, H., Wong-Foy, A., McCoy, B., Kim, S., Eckerle, J., Low, T. (2011). From boots to buoys: Promises and challenges of dielectric elastomer energy harvesting. *Conference on Electroactive Polymer Actuators and Devices (EAPAD) 2011*, San Diego, USA, March 7–10.

Kornbluh, R.D., Pelrine, R., Prahlad, H., Wong-Foy, A., McCoy, B., Kim, S., Eckerle, J., Low, T. (2012). Dielectric elastomers: Stretching the capabilities of energy harvesting. *Mrs Bulletin*, 37(3), 246–253.

Kuang, Z., Chen, Y., Lu, Y., Liu, L., Hu, S., Wen, S., Mao, Y., Zhang, L. (2015). Fabrication of highly oriented hexagonal boron nitride nanosheet/elastomer nanocomposites with high thermal conductivity. *Small (Weinheim an der Bergstrasse, Germany)*, 11(14), 1655–1659.

Lallart, M., Cottinet, P.-J., Guyomar, D., Lebrun, L. (2012). Electrostrictive polymers for mechanical energy harvesting. *Journal of Polymer Science Part B – Polymer Physics*, 50(8), 523–535.

Lehmann, W., Skupin, H., Tolksdorf, C., Gebhard, E., Zentel, R., Kruger, P., Losche, M., Kremer, F. (2001). Giant lateral electrostriction in ferroelectric liquid-crystalline elastomers. *Nature*, 410(6827), 447–450.

Li, J.Y. (2003). Exchange coupling in P(VDF-TrFE) copolymer based all-organic composites with giant electrostriction. *Physical Review Letters*, 90(21), 217601.

Lipomi, D.J., Vosgueritchian, M., Tee, B.C.K., Hellstrom, S.L., Lee, J.A., Fox, C.H., Bao, Z. (2011). Skin-like pressure and strain sensors based on transparent elastic films of carbon nanotubes. *Nature Nanotechnology*, 6(12), 788–792.

Luna, A., Yuan, J., Neri, W., Zakri, C., Poulin, P., Colin, A. (2015). Giant permittivity polymer nanocomposites obtained by curing a direct emulsion. *Langmuir*, 31(44), 12231–12239.

Luna, A., Pruvost, M., Yuan, J.K., Zakri, C., Neri, W., Monteux, C., Poulin, P., Colin, A. (2017). Giant electrostrictive response and piezoresistivity of emulsion templated nanocomposites. *Langmuir*, 33(18), 4528–4536.

Minh Quyen, L., Capsal, J.-F., Galineau, J., Ganet, F., Yin, X., Yang, M., Chateaux, J.-F., Renaud, L., Malhaire, C., Cottinet, P.-J., Liang, R. (2015). All-organic electrostrictive polymer composites with low driving electrical voltages for micro-fluidic pump applications. *Scientific Reports*, 5, 11814.

Molberg, M., Crespy, D., Rupper, P., Nueesch, F., Manson, J.-A.E., Loewe, C., Opris, D.M. (2010). High breakdown field dielectric elastomer actuators using encapsulated polyaniline as high dielectric constant filler. *Advanced Functional Materials*, 20(19), 3280–3291.

Nesser, H., Debeda, H., Yuan, J.K., Colin, A., Poulin, P., Dufour, I., Ayela, C. (2018). All-organic microelectromechanical systems integrating electrostrictive nanocomposite for mechanical energy harvesting. *Nano Energy*, 44, 1–6.

O'Halloran, A., O'Malley, F., McHugh, P. (2008). A review on dielectric elastomer actuators, technology, applications, and challenges. *Journal of Applied Physics*, 104(7), 071101.

Park, C., Kang, J.H., Harrison, J.S., Costen, R.C., Lowther, S.E. (2008). Actuating single wall carbon nanotube-polymer composites: Intrinsic unimorphs. *Advanced Materials*, 20(11), 2074–2079.

Pelrine, R., Kornbluh, R.D., Pei, Q.B., Joseph, J. (2000). High-speed electrically actuated elastomers with strain greater than 100%. *Science*, 287(5454), 836–839.

Pelrine, R., Kornbluh, R.D., Eckerle, J., Jeuck, P., Oh, S.J., Pei, Q.B., Stanford, S. (2001). Dielectric elastomers: Generator mode fundamentals and applications, *Smart Structures and Materials 2001 Conference*, Newport Beach, USA, March 5–8.

Romasanta, L.J., Lopez-Manchado, M.A., Verdejo, R. (2015). Increasing the performance of dielectric elastomer actuators: A review from the materials perspective. *Progress in Polymer Science*, 51, 188–211.

Su, J., Harrison, J.S., St Clair, T.L., Bar-Cohen, Y., Leary, S. (2000). Electrostrictive graft elastomers and applications. In *Materials Research Society Symposium Proceedings, Volume 600, Electroactive Polymers*, Zhang, Q.M., Furukawa, T., Bar-Cohen, Y., Scheinbeim, Y. (eds). Cambridge University Press, Cambridge.

Su, J., Costen, R.C., Harrison, J.S. (2002). Performance evaluation of bending actuators made from electrostrictive graft elastomers. *Smart Structures and Materials 2002 Conference*, San Diego, USA, March 18–21.

Vijaya, M.S. (2013). *Piezoelectric Materials and Devices: Applications in Engineering and Medical Sciences*. CRC Press, Boca Raton.

Wang, Y.Q., Sun, C.J., Zhou, E., Su, J. (2004). Deformation mechanisms of electrostrictive graft elastomer. *Smart Materials & Structures*, 13(6), 1407–1413.

Wang, T.S., Farajollahi, M., Choi, Y.S., Lin, I.T., Marshall, J.E., Thompson, N.M., Kar-Narayan, S., Madden, J.D.W., Smoukov, S.K. (2016). Electroactive polymers for sensing. *Interface Focus*, 6(4), 20160026.

Wongtimnoi, K. (2011). Polyurethanes électrostricitifs et nanocomposites : caractérisation et analyse des mécanismes de couplage électromécaniques. PhD Thesis, Institut National des Sciences Appliquées de Lyon.

Wongtimnoi, K., Guiffard, B., Bogner–Van de Moortele, A., Seveyrat, L., Gauthier, C., Cavaille, J.Y. (2011). Improvement of electrostrictive properties of a polyether-based polyurethane elastomer filled with conductive carbon black. *Composites Science and Technology*, 71(6), 885–892.

Wongtimnoi, K., Guiffard, B., Bogner–Van de Moortele, A., Seveyrat, L., Cavaille, J.Y. (2013). Electrostrictive thermoplastic polyurethane-based nanocomposites filled with carboxyl-functionalized multi-walled carbon nanotubes (MWCNT-COOH): Properties and improvement of electromechanical activity. *Composites Science and Technology*, 85, 23–28.

Xia, F., Cheng, Z.Y., Xu, H.S., Li, H.F., Zhang, Q.M., Kavarnos, G.J., Ting, R.Y., Abdul-Sedat, G., Belfield, K.D. (2002). High electromechanical responses in a poly (vinylidene fluoride-trifluoroethylene-chlorofluoroethylene) terpolymer. *Advanced Materials*, 14(21), 1574–1577.

Yang, Y., Gao, Z.-S., Yang, M., Zheng, M.-S., Wang, D.-R., Zha, J.-W., Wen, Y.-Q., Dang, Z.-M. (2019). Enhanced energy conversion efficiency in the surface modified $BaTiO_3$ nanoparticles/polyurethane nanocomposites for potential dielectric elastomer generators. *Nano Energy*, 59, 363–371.

Yin, G.L., Yang, Y., Song, F.L., Renard, C., Dang, Z.M., Shi, C.Y., Wang, D.R. (2017). Dielectric elastomer generator with improved energy density and conversion efficiency based on polyurethane composites. *ACS Applied Materials & Interfaces*, 9(6), 5237–5243.

Yuan, J. (2017). Percolation of carbon nanomaterials for high-k polymer nanocomposites. *Chinese Chemical Letters*, 28(11), 2036–2044.

Yuan, J., Yao, S.-H., Dang, Z.-M., Sylvestre, A., Genestoux, M., Bai, J. (2011). Giant dielectric permittivity nanocomposites: Realizing true potential of pristine carbon nanotubes in polyvinylidene fluoride matrix through an enhanced interfacial interaction. *Journal of Physical Chemistry C*, 115(13), 5515–5521.

Yuan, J., Yao, S., Li, W., Sylvestre, A., Bai, J. (2014). Vertically aligned carbon nanotube arrays on SiC microplatelets: A high figure-of-merit strategy for achieving large dielectric constant and low loss in polymer composites. *Journal of Physical Chemistry C*, 118(40), 22975–22983.

Yuan, J., Luna, A., Neri, W., Zakri, C., Schilling, T., Colin, A., Poulin, P. (2015). Graphene liquid crystal retarded percolation for new high-k materials. *Nature Communications*, 6, 8700.

Yuan, J., Yao, S., Li, W., Sylvestre, A., Bai, J. (2017). Anisotropic percolation of SiC–carbon nanotube hybrids: A new route towards thermally conductive high-k polymer composites. *Journal of Physical Chemistry C*, 121(22), 12063–12070.

Yuan, J., Luna, A., Neri, R., Zakri, C., Colin, A., Poulin, P. (2018). Giant electrostriction of soft nanocomposites based on liquid crystalline graphene. *ACS Nano*, 12(2), 1688–1695.

Yuan, J., Neri, W., Zakri, C., Merzeau, P., Kratz, K., Lendlein, A., Poulin, P. (2019). Shape memory nanocomposite fibers for untethered high-energy microengines. *Science*, 365(6449), 155–158.

Zhang, Q.M., Su, J., Kim, C.H., Ting, R., Capps, R. (1997). An experimental investigation of electromechanical responses in a polyurethane elastomer. *Journal of Applied Physics*, 81(6), 2770–2776.

Zhang, Q.M., Bharti, V., Zhao, X. (1998). Giant electrostriction and relaxor ferroelectric behavior in electron-irradiated poly(vinylidene fluoride-trifluoroethylene) copolymer. *Science*, 280(5372), 2101–2104.

Zhang, Q.M., Li, H.F., Poh, M., Xia, F., Cheng, Z.Y., Xu, H.S., Huang, C. (2002). An all-organic composite actuator material with a high dielectric constant. *Nature*, 419(6904), 284–287.

List of Authors

Jinbo BAI
CentraleSupélec
Paris-Saclay University
Gif-sur-Yvette
France

Fahmi BEDOUI
Université de technologie de Compiègne
Sorbonne University
France

Florentin BERTHET
Institut Clément Ader
University of Toulouse
France

Adoté Sitou BLIVI
Université de technologie de Compiègne
Sorbonne University
France

France CHABERT
ENIT – INPT
University of Toulouse
France

Stéphane CUYNET
Institut Jean Lamour
University of Lorraine
Nancy
France

Karl DELBÉ
ENIT – INPT
University of Toulouse
France

Jean DENAPE
ENIT – INPT
University of Toulouse
France

Fabrice DETREZ
MSME
Gustave Eiffel University
Marne-la-Vallée
France

Marie DOUMENG
ENIT – INPT
and
Institut Clément Ader
University of Toulouse
France

Benhui FAN
CentraleSupélec
Paris-Saclay University
Gif-sur-Yvette
France

Sébastien FONTANA
Institut Jean Lamour
University of Lorraine
Nancy
France

Jean-François GÉRARD
INSA
Lyon
France

Ann-Lenaig HAMON
CentraleSupélec
Paris-Saclay University
Gif-sur-Yvette
France

Delong HE
CentraleSupélec
Paris-Saclay University
Gif-sur-Yvette
France

Djimédo KONDO
Institut Jean Le Rond d'Alembert
Sorbonne University
Paris
France

Adrien LETOFFE
Institut Jean Lamour
University of Lorraine
Nancy
France

Yu LIU
CentraleSupélec
Paris-Saclay University
Gif-sur-Yvette
France

Xiaoxin LU
Shenzhen Institutes of Advanced Technology
Chinese Academy of Sciences
China
and
MSME
Gustave Eiffel University
Marne-la-Vallée
and
CentraleSupélec
Paris-Saclay University
Gif-sur-Yvette
France

Olivier MARSAN
CIRIMAT – ENSIACET – INPT
University of Toulouse
France

Angel MORA
CICY
Mérida
Mexico

Marc PONÇOT
Institut Jean Lamour
University of Lorraine
Nancy
France

Lucie SPEYER
Institut Jean Lamour
University of Lorraine
Nancy
France

Jinkai YUAN
Centre de Recherche Paul Pascal
CNRS – University of Bordeaux
Pessac
France

Shenghong YAO
Centre de Recherche Paul Pascal
CNRS – University of Bordeaux
Pessac
France

Julien YVONNET
MSME
Gustave Eiffel University
Marne-la-Vallée
France

Index

A, C

D

E

F, G

H, I

L, M

N

P

R, S

T, U, X